The Origins and Use of the Potter's Wheel in Ancient Egypt

S. K. Doherty

Archaeopress Egyptology 7

Archaeopress
First and Second Floors
13-14 Market Square
Bicester OX26 6AD

www.archaeopress.com

ISBN 978 1 78491 060 0
ISBN 978 1 78491 061 7 (e-Pdf)

© Archaeopress and S K Doherty 2015

QR Code:

http://bit.do/potterswheel

All rights reserved. No part of this book may be reproduced, stored in retrieval system, or transmitted, in any form or by any means, electronic, mechanical, photocopying or otherwise, without the prior written permission of the copyright owners.

This book is available direct from Archaeopress or from our website www.archaeopress.com

To the memory of Amanda, who always thought that even the smallest achievements should be properly recorded.

Also to the memory of my cousin Ellen Doherty, taken from our family too soon and who shared my deep love of Africa.

Contents

List of Figures ii
List of Tables v
Acknowledgements vii

Chapters

Introduction 1

Seeking the Potter's Wheel 4

Ancient Sources for the potter's wheel 23

Inventing the potter's wheel 38

How did the Potter's Wheel come to Egypt? 55

Detecting the Use of the Potter's Wheel in Egyptian Pottery 70

The Spread of the Potter's Wheel from Royal to Domestic Contexts 92

Conclusion 106

References 112

Appendices

Appendix I 128

Appendix II 132

Appendix III 134

Appendix IV 138

Potter's Wheel Videos availabe at https://www.youtube.com/user/Ramessesmissy/feed or Archaeopress Open Access site, http://www.archaeopress.com/ArchaeopressShop/Public/defaultAll.asp?OpenAccess=Y&intro=true

List of Figures

Figure 2.1: Different Types of Potter's Wheels with French Mistranslations 5

Figure 2.2: Tomb of Ty showing potter 6

Figure 2.3: An example of an s-shaped crack, indicative of thrown pottery 7

Figure 2.4: The Twist Reverse Twist Drill 7

Figure 2.5: Manufacturing Meidum bowls over a mould or former 8

Figure 2.6: An example of a CD7 bowl, 4th dynasty, Giza 8

Figure 2.7: An example of the pierced wheel bearing from Tel Dalit 15

Figure 2.8: Japanese potter's wheel with socketed disk. Arrangement of basalt bearings 15

Figure 2.9: Clay wheelhead from pottery quarter at Ur 16

Figure 2.10: The remains of the potter's workshop 17

Figure 2.11: The Abu Sir clay wheel-head. Verner's interpretation of how the wheel was set up 18

Figure 2.12: The reconstructed potter's wheel 18

Figure 2.13: The British Museum Collection of unprovenanced Egyptian Potter's Wheel bearings 21

Figure 3.1: Possible potter's wheel scene from the rock cut tomb of Nebemakhet, Giza 23

Figure 3.2: Potter's Workshop from tomb of Ty 24

Figure 3.3: Stele of Prince Wepemnefret by Norman de Garis Davies 25

Figure 3.4: Tomb of Khentika from Saqqara, in the cemetery of Pharoah Teti 25

Figure 3.5: The loose block of a seated potter working on his potter's wheel 26

Figure 3.6: The tomb of Bakt III pottery making scene 27

Figure 3.7: The pottery workshop scene from the tomb of Amenemhat 28

Figure 3.8: The pottery workshop of Khnumhotep III at Beni Hasan 29

Figure 3.9: The potter representation from the pottery scene from tomb of Nomarch Djeutihotep 30

Figure 3.10: The pottery workshop scene in the tomb of Djeutihotep 30

Figure 3.11: The potters from the tomb of Horemkawef, Hierakonpolis 31

Figure 3.12: Pottery workshop of Kenamun (TT 93), Thebes 32

Figure 3.13: Servant Statuette of Potter 32

Figure 3.14: A close up of the potter's wheel in Gemniemhat's tomb at Saqqara 32

Figure 3.15: Wooden model from the tomb of Gemniemhat at Saqqara 33

Figure 3.16: Relief from the tomb of the 5th dynasty Vizier Ptahshepses 35

Figure 3.17: Section of papyrus from the archive of the Raneferef's mortuary temple 35

Figure 3.18: An example of a handmade coil built beer jar 35

Figure 3.19: The Pyramid text representations of potters 35

Figure 4.1: The Chaîne Opératoire approach 39

Figure 4.2: An example of Petrie's Black topped ware 42

Figure 4.3: An example of a Bevelled rim bowl 42

Figure 4.4: Representation of a shrine on the top of the Uruk Vase 42

Figure 4.5: Map of the Near East and Egypt, showing keys sites mentioned in the text............................ 43

Figure 4.6: Polychrome handbuilt pottery.. 44

Figure 4.7: The plan of the city of Uruk-Warka.. 45

Figure 4.8: The Twist Reverse Twist Drill 48

Figure 4.9: Door Socket made of Quartzite found near to the temple revetment at Hierakonpolis 48

Figure 4.10: Map of the Town of Hierakonpolis 49

Figure 4.11: Map of Egypt showing basalt outcrops 50

Figure 4.12: Niuserre upper temple, Abu Sir 5th Dynasty... 51

Figure 4.13: Examples of V-shaped bowls, made by arranging coils of clay.. 52

Figure 4.14: The Chaine Operatoire of the v-rimmed bowl.. 53

Figure 5.1: Ceramics from Stratum Ia in Buto. 56

Figure 5.2: The Complete wall from the storeroom of the tomb of Ty.. 58

Figure 5.3: Example of a man making pottery using the hammer and anvil technique 59

Figure 5.4: Evidence for social status of the potter at the wheel displaying prominent ribs 59

Figure 5.5: Experimental reconstruction of the pit kilns located at HK11 C Square B4NW........................ 62

Figure 5.6: The Fire dog features from Hierakonpolis square A6, HK11 C.. 63

Figure 5.7: The screen kiln at el Mahasna 63

Figure 5.8: The Assistant Potter in the tomb of Ty in front of the kiln 64

Figure 5.9: The multi-period pottery workshop at Ain Asil 64

Figure 5.10: Hazor pottery mask and wheel bearing in situ 65

Figure 5.11: Plan and section of cave 4034 at Lachish 66

Figure 5.12: The Miniature vessel dump outside Sneferu's Meidum pyramid 67

Figure 5.13: Examples of miniature vessels from Meidum.. 67

Figure 5.14: The Meidum Pyramid foundation deposit.. 68

Figure 6.1: The wheel-thrown pot and the hand-built coil pot at leather hard stage............................... 72

Figure 6.2: X-rays of the coil hand-built experimental pot and electric wheel-thrown pot........................ 72

Figure 6.3: Xeroradiograph of three miniature vessels 72

Figure 6.4: Indications of thrown pottery... 73

Figure 6.5: Iron Oxide Spangles being added to the clay during the wedging process 73

Figure 6.6: Coils clearly visible in the base of this wavy handled jar c.3200 B.C. 74

Figure 6.7: The rilling marks created by the fingers of the potter 74

Figure 6.8: Wavy handled jar. Constructed using coils on flat support.. 79

Figure 6.9: Miniature Vessel from Abydos. Thrown on a potter's wheel 79

Figure 6.10: The characteristic marks of wheel-throwing 81

Figure 6.11: BM32622 82

Figure 6.12: BM32622 82

Figure 6.13: The newly cured concrete potter's wheel bearings... 83

Figure 6.14: The original sketch of the potter's wheelhead found in the mortuary temple 84

Figure 6.15: Attaching the wheelhead to the concrete wheel bearing using coils of clay......................... 85

Figure 6.16: The reconstructed potter's wheel 85

Figure 6.17: The author has finished centring the lump of clay on the reconstructed ancient wheel 86

Figure 6.18: The pottery tools found in the potter's workshop at Lachish..................................86

Figure 6.19: Author reapplying lubricant to the concrete potter's wheel replica 87

Figure 6.20: The carved and honed granite replica potter's wheel bearings 87

Figure 6.21: The granite wheel bearings set up 87

Figure 6.22: The outside of the replica pot. The outside of the archaeological miniature vessel.............88

Figure 6.23: The inside of the replica pot. The inside of the archaeological miniature vessel...................89

Figure 6.24: The bases of the replica pots. The bases of the archaeological miniature vessel 89

Figure 6.25: Examples of V-shaped bowls, made by arranging coils of clay................................90

Figure 6.26: Internal view of replicated V-rim vessel (unfired) 90

Figure 6.27: The smoothed outer edge of the replicated V-rim vessel 91

Figure 7.1: The Statue of Djoser's ka from his serdab at Saqqara 93

Figure 7.2: A dummy stone model vase made of calcite. Meidum vessel red slipped pottery 94

Figure 7.3: Model vessels made of Calcite from Giza tomb G 7440 Z, 4th dynasty...................95

Figure 7.4: 4th dynasty miniature vessels from Meidum 96

Figure 7.5: Shape comparison of Predynastic (Naqada I-II) basalt stone vessels 98

Figure 7.6: Built area of the mortuary temple of Old Kingdom Kings 100

Figure 7.7: G. A. Reisner's 1930s pottery spoil heap still visible to the south of Khafre's pyramid..........100

Figure 7.8: Large conical *bedjᶜ* bread mould manufactured around a conical former Wodzińska..........101

Figure 7.9: The experimental wheel set up with pre-prepared cones of clay 102

Figure 7.10: The rilling marks are quite clearly discernible in this Meidum vessel sherd from Buhen 102

Figure 7.11: 6th dynasty bowl with spouted rim, from Saqqara SQ98-507 Type 598 103

Figure 7.12: Close up detail of a Meidum bowl rim sherd showing the rilling marks similar to Figure....103

Figure 7.13: Three views of the same CD7 vessel AW1275, from Heit el Ghurob, Giza 103

Figure 7.14: Drawing of CD7 bowl made of Nile Clay. Example of Meidum bowl from Giza 104

List of Tables

Table 2.1: Table of Provenanced Potter's Wheel Bearings and Wheelheads in Egypt and the Near East .. 10

Table 2.2: Selected Potter's Wheels in World Museum Collections 19

Table 3.1: Wooden models of Potter's workshops and their details 33

Table 5.1: Showing the percentages of different professions mentioned in Papyrus BM 10068 60

Table 6.1: Manufacturing Marks Criterion 75

Table 6.2: Examples of Macroscopic Details for Coiling in Museum Pottery Collections 80

Table 6.3: Table of Macroscopic Details for Wheel throwing in Museum Collections 81

Acknowledgements

This book is the ultimate product of three years' worth of PhD research at Cardiff University. It is quite unusual to finish a thesis (let alone a book!) on time without considerable support from others along the way. I wish to thank in particular Prof. Paul Nicholson for his inspiring and generous discussions, which helped to shape my thinking, combined with regular feedback, and many wise words of advice regarding my research. Grateful thanks are due to Prof. Ian Freestone for encouraging my experimental research and supporting my scientific endeavours. Thank you to Dr. Alan Lane for stepping in as my 2nd supervisor during my third year. Sincere thanks to Prof. Alan Davies and Stephen P. Meade and the technicians of ENGIN who helped me design, make, and improve the concrete bearings. Grateful acknowledgements to the Cyril Fox Fund for providing the necessary financial aid to manufacture the granite wheel bearings. Much appreciation is due to the staff of Archaeopress for enabling this book to be published.

I owe a huge amount of thanks to my aunt Joan Doherty and the Clay Hill potters who let me take over Clay Hill Pottery for my experiments. Joan thanks for sharing all your expertise with me and helping me learn new potting skills. I am extremely grateful to the staff of the British Museum, the Petrie Museum of Egyptian Archaeology, UCL, Cyfarthfa Castle Museum, Merthyr Tydfil, the Ashmolean Museum, Oxford for allowing me to view and photograph their pottery collections. Especial thanks to Xavier Droux at the Ashmolean Museum, Oxford, Tracey Golding and Alice Stevenson of the Petrie Museum for all their help. Thanks to Jimmy Peake for helping me to use the X-Ray at Cardiff University and to Jane Henderson for permitting me to do so. Sincere thanks to Anna Wodzińska and Teodozja Rzeuska for inviting me to contribute to the Old Kingdom Pottery Workshop: Chapter 2 in Warsaw, which proved to be very thought provoking, particularly in relation to Giza and Saqqara pottery types.

My outstanding colleagues of the Gurob Harem Palace Project directed by Dr. Ian Shaw and expertly organised by Jan Picton and Ivor Pridden are too many to mention, but thank you all for making my fieldwork experience in Egypt so wonderful. The Postgraduate Quality Committee, SHARE and the Cardiff Alumni Award, and the Carlsberg Foundation are acknowledged for kindly funding my fieldwork at Gurob and ethnographic work at El-Nazla pottery. Grateful thanks are due to fellow Gurobite Dr. Tine Bagh for sending me photos (via Paul & Ivor) of the potter's wheel bearing from the Ny Carlsberg Glyptotech and to Guy Lecuyot for the Saqqara example. Many thanks to Duncan and Matt of Cardiff Metropolitan University Ceramics department who were so welcoming and allowed me to take advantage of their departments' excellent facilities.

Much love and grateful thanks to Lyn, Max & Raminta for offering me a place to stay in London, providing excellent dinners, combined with amazing Egyptology chats. To Caroline Doherty, thank you for helping me put everything into perspective. Waves of appreciation to my fellow Cardiff PhD Postgraduates who regularly provided me with a sympathetic shoulder and interesting insights into the sometimes-intensive PhD experience. I am indebted to my colleagues Carolyn and Marylyn at the MBI Al Jaber Foundation for encouraging me to publish this book. Considerable (and ongoing!) love and thanks to Lorna and James Doherty for all their support and wonderful advice. My PhD thesis and now this book was conceived when my uncle Jack Doherty, a Potter asked me casually over a pint, "So what is the history of the Potter's Wheel then?" Well Jack, three years later, this is the result!

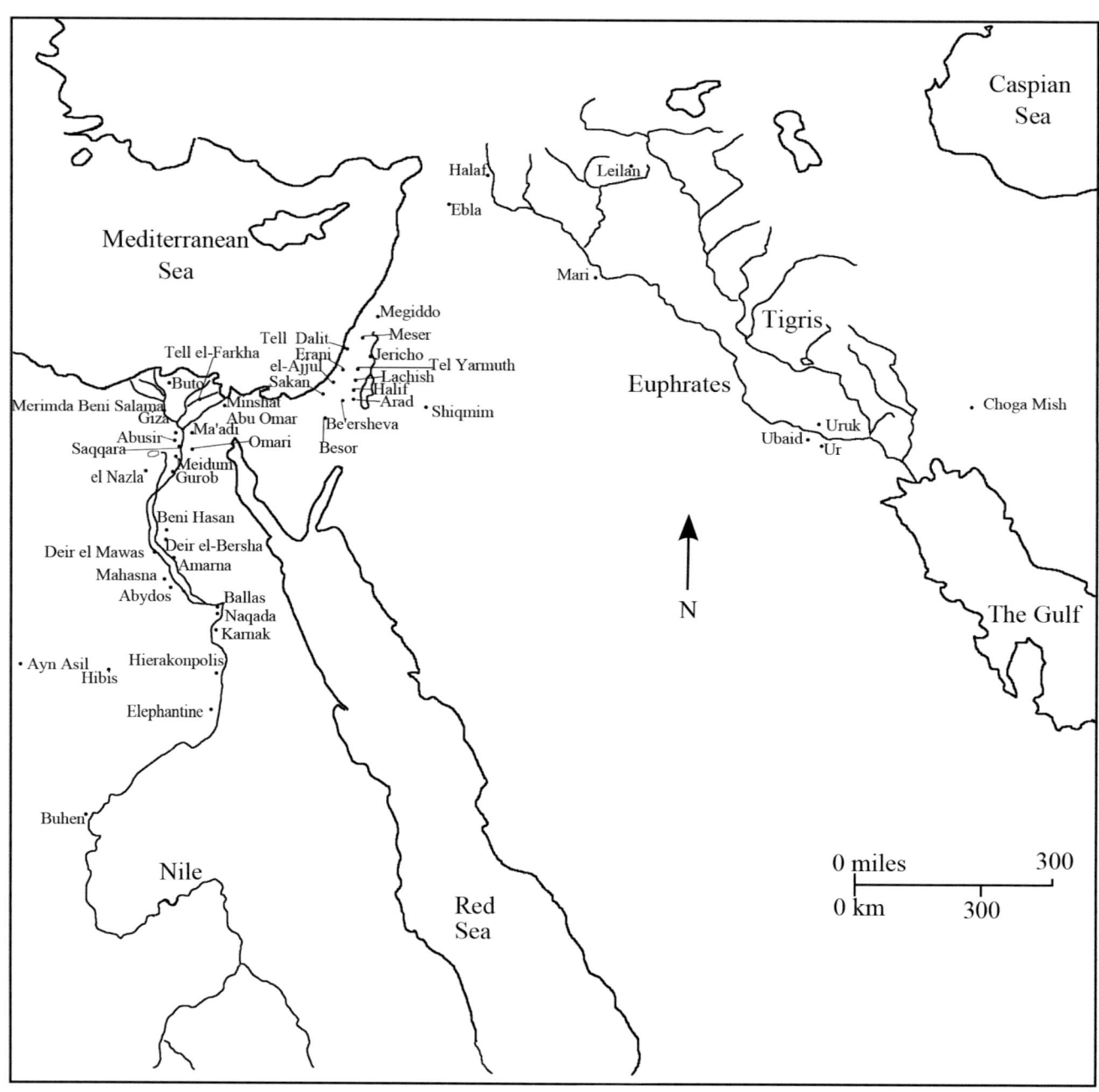

Map of the Near East and Egypt, showing keys sites mentioned in the text. S. Doherty

Cal. BC	Petrie's Phases (Petrie 1901, p. 4-12)	Petrie's Sequence Dates	Period-Lower Egypt (Hassan, 1985, pp. 95-6, fig 2)	Period-Upper Egypt (Kaiser, 1957, pp. 67-77)	Period (Hendrickx 1996, pp. 63-64)	Key Kings	Dynasties	Near East Terms (Dessel & Joffe, 2000, p. 38)
c1550-1069	New Kingdom		New Kingdom	New Kingdom	New Kingdom	Seti I Ramesses II	18-20	LATE BRONZE AGE 1550 - 1200 BC
c1700-1550	Second Intermediate Period		Second Intermediate Period	Second Intermediate Period	Second Intermediate Period	Hyksos in Delta Pharaohs ruled Thebes	13-17	MIDDLE BRONZE AGE IIB/C (1750-1500 BC)
c.2025-1700	Middle Kingdom		Middle Kingdom	Middle Kingdom	Middle Kingdom	Amenemhat III Nimaatre	10-12	MIDDLE BRONZE AGE IIA (2000-1750 BC)
c.2181-2025	First Intermediate Period		First Intermediate Period	First Intermediate Period	First Intermediate Period	Neferirkare	7-10 (11th also in south)	MIDDLE BRONZE AGE I (2100-2000 BC) EARLY BRONZE AGE IV (2100 BC)
c. 2686-2181	Old Kingdom		Old Kingdom	Old Kingdom	Old Kingdom	Sneferu, Khufu, Niuserre, Unas, Pepi I & II	3-6	EARLY BRONZE AGE III/EARLY DYNASTIC II c2750 BC EARLY DYNASTIC III c2600
c. 2900		75-82			Naqada IIID	Semerkhet/Qa'a	2	EARLY DYNASTIC/ EARLY BRONZE AGE II
c. 3000-2900		75-82		Naqada III c3	Naqada IIIC2	Djed-Adjib	2	URUK/PROTOLITERATE/ EB II
c. 3100-3000	Semainean (until 3050 Cal. BC)	63-76		Naqada III c1-2	Naqada IIIC1	Narmer-Djer	1	URUK/PROTOLITERATE EARLY BRONZE AGE II
c. 3300-3100	Semainean	63-76		Naqada III a1-IIIb2	Naqada IIIA1-IIIB	Scorpion I-Iri-Hor/Ka	0	URUK/PROTOLITERATE EARLY BRONZE AGE I
c. 3650-3300	Gerzean	38-62	Buto (c-d) Ma'adi (a-b)	Naqada IIc-IIIa1	Naqada IIC-IID2			UBAID CHALCOLITHIC 4000-3000 BC
c. 3900-3650	Amratian	31-37	Ma'adi	Naqada Ia-IIb	Naqada IA-IIB			UBAID- c5000-3500 BC
c. 5200-3900	Badarian		Faiyum Neolithic 5200-4000 BC	Badarian c 4400-3900 BC	Badarian			UBAID c 5000-3500 BC

CHRONOLOGY. DATES UNCERTAIN PRIOR TO 664 B.C.

Chapter 1:

Introduction

Despite many years work by scholars on the technology of pottery production, it is perhaps surprising that the origins of the potter's wheel in Egypt has yet to be determined. This present project seeks to rectify this situation by (1) determining when the potter's wheel was introduced into Egypt, (2) establishing in what contexts wheel-made pottery occurs, and (3) considering the reasons why the Egyptians introduced the wheel when a well-established hand-made pottery industry already existed. To date, research has tended to focus on the decoration and function of the pot rather than on the manufacturing methods used. In the early part of the twentieth century, mention of the potter's wheel was often a brief comment indicating that the wheel seemed to be in use rather than discussion on how it came to be used as a technology or how the use of the wheel was reflected on the pottery (Reisner, 1923; Petrie, 1925, p. 57).

The reasons why the potter's wheel came to Egypt have not yet been sufficiently discussed, nor has the first use of the wheel in Egypt been completely ascertained, yet the potter's wheel is arguably the most significant machine introduced into Egypt during the Old Kingdom, second only perhaps to the lever. Most ancient inventions were inspired by shapes noted in the natural world. Wheels do not exist in nature, and so can be viewed entirely as a human-inspired invention. The impact of this innovation would not just have affected the Egyptian potters themselves through the learning of a new skill but it also signalled the beginnings of a more complex and technologically advanced nation. The links between the potter's wheel and the rise of elite-sponsored specialisation have not yet been examined. It is through a thorough analysis of all available sources, such as manufacturing marks on pottery, provenanced potter's wheels, and depictions of potters in art and text that the origins of the potter's wheel can begin to be understood. Through examining manufacturing marks on pottery and determining which are characteristic of wheel-made wares by comparing these marks with experimental examples, it is hoped that a more complete view can be gained about when and in what manner the Egyptians were manufacturing their pottery vessels on the wheel.

There are terminological problems amongst the literature relating to the potter's wheel. Scholars are uncertain whether the wheel bearings discovered on excavation sites or depicted on tomb walls should be termed a turntable or a potter's wheel. There is also uncertainty about whether these bearings were actually capable of producing thrown pottery or were instead being used as an aid for rotating a vessel during handbuilding. As a result, a variety of terms exist and researchers (Arnold, 1993, pp. 41-3; Edwards & Jacobs, 1986, pp. 55-6; Rieth, 1960, p. 20) do not seem to agree on whether these bearings should be termed potter's wheel, fast simple (low) wheel (Holthoer, 1977, p. 31), low wheel, slow (simple) wheel (Rice, 1987, pp. 132-4), potter's stand, turntable (Edwards & Jacobs 1986, pp. 55-56;1987), *Töpferscheibe* (Arnold, 1976; Faltings, 1989, p. 137), *tour, tournage* or *tournette* (Childe 1954, pp. 196-197; Soukiassian *et al.* 1990). In addition, one of the major debates regarding the use of the potter's wheel focuses on whether a centrifugal force[1] of sufficient rotations per minute (r.p.m.) can be achieved to throw a pot c.50-150 r.p.m. (Rye, 1981, p. 74)[1] or whether it could be achieved at lower speeds contra to Edwards and Jacobs (1986, pp. 55-56;1987).

Another debate concerns whether vessels were in fact "rotated"[2] on the "wheel" as part of the finishing process, with the resulting concentric rings or rilling marks created by "Rotative Kinetic Energy" or whether this "RKE" made the vessel appear as though it was thrown (Roux, 2003, p. 23; Roux & de Miroschedji, 2009). Dorothea Arnold (1993, p. 42) notes that the term "turning" is sometimes applied to pots that have been slowly rotated on a slow (hand-spun) wheel, and suggests that a better term to use would be "rotational assisted device" or turntable. The use of the terms concentric rings and rilling are equally applied to a pot that has been rotated or thrown, or a combination of the two, and this can often lead to confusion. Some pots are described as "partially rotated" implying that only a particular section of the vessel was formed on a wheel, often the rim of the vessel (Arnold, 1993, p. 36; Wodzińska, 2009c, p. 25) or "wheel shaped" (Roux 2003, p. 3) meaning that the wheel was used to thin down or shape already roughly coiled vessels. These terminology problems will be further addressed in Chapter 2 and in experiments in Chapter 6.

The scope of Chapters 2 and 3 is to review the known evidence relating to when the potter's wheel was first utilised in ancient Egypt. The archaeological literature will be consulted to determine the present state of knowledge, and with any problems, terminological contradictions, errors, or misnomers highlighted for further examination

---

[1] Not to be confused with the term centripetal force. Centripetal force, from the Latin for "*centre seeking*" is a centre seeking force through which the force is always directed toward the centre of the circle. Without this force, an object will simply continue moving in a straight-line motion. By contrast, centrifugal force, from the Latin for "*central fleeing*," relates to moving or direction outward from the centre, this is the opposite of centripetal force. Centrifugal force is occurring within the clay when the potter's wheel is spun sufficiently fast, the clay is directed outward from the centre of the wheel.
[2] Archaeologists (e.g. Arnold 1993) sometimes use the misnomer "*turned*" to signify rotated, whereas potters use the term to indicate the scraping or shaving off any excess clay.

later in this thesis. Chapter 2 will guide the reader through many of the known excavated potters' wheels, whether provenanced or not, in the Near East and Egypt. In Chapter 3, an analysis of the known tomb art depicting potter's wheels and workshops, tomb models of workshops and limestone statuettes in Egypt only, as research to date has not revealed relevant tomb art from the Near East will be undertaken. Finally, Chapter 3 will describe the known ancient textual and written sources relating to potters to provide a broad overview of all possible sources before they can be thoroughly analysed.

Chapter 4 will consider whether the potter's wheel was used differently in Egypt than in other areas of the Old World. It is suggested that potters in the Near East did not initially utilise the potter's wheel for throwing vessels, whereas the Egyptians did. By understanding how the pottery industries developed within the Ancient Near East and Egypt it is hoped that the underlying social and economic structures can be understood. If both areas had similar pottery industries based upon workshops, kilns and wheel production run by specialist potters perhaps being instigated or organised through elite-sponsorship, then it is likely that the two pottery industries developed from the same model. Inventions such as the potter's wheel may have been transferred to Egypt from Near Eastern centres in a form of elite technological exchanges from one court to another as part of diplomatic relations. Evidence for such exchanges has been well documented in terms of art styles, foreign pottery influences (Faltings, 1998a, 1998b; Von der Way, 1992), foreign imports (Oren & Yekutieli, 1992, pp. 361-384) and the Egyptian colonisation of Canaan (Brandl, 1992, pp. 441-448).

The Egyptian hierarchical structuring of Dynastic times is thought to have been quite rigid and controlling of the lower status members of society (Shaw, 2004, pp. 12-24) but is this reflected upon the status of Egyptian potters? The status of the potter will be determined through study of the representation of potters in art e.g. tomb wall scenes, textual evidence such as the *Satire of the Trades*,[3] archaeological remains such as pottery workshop sites, and comparisons with modern ethnographic studies of potters. Any change in the status of potters could be related to broader socio-political changes within the Egyptian state, and could be a wider ranging phenomenon occurring concurrently in contemporary societies in the Near East. Through extensive reading of technological theory and gender theory and applying this to the Egyptian model, it is hoped to trace the development of the invention of the potter's wheel to the production of pottery using the potter's wheel. Pottery made by hand is often thought to be the realm of women, but when the wheel begins to be used, men tend to be the main potters (Vincentelli, 2003). Through the application of gender theory and ethnographic study the role of Egyptian men and women in pottery production will be assessed in Chapter 4.

The stone wheel bearings which form the main moving component of the potter's wheel were usually made of basalt or granite (see Table 2.2, Chapter 2; Hope, 1981; Powell, 1995), two of the hardest stones to quarry, hew, hone (7 on the Mohs scale, Tabor, 1954, p. 251) and procure as they are often sourced in far-flung, hazardous locations. Therefore, quarrying expeditions would require much elite-instigated forethought and organisation (Harell & Brown, 1995; Klemm & Klemm, 1993; Mallory-Greenough, Greenough, & Owen, 1999). Chapter 4 will assess the significance of the use of basalt and granite, which during the Old Kingdom were normally restricted to the production of elite royal funerary items such as vases (Mallory-Greenough *et al.* 1999), mortuary pyramid temple floors (Hoffmeir 1993, p. 117; Mallory-Greenough *et al.* 2000) boundary or tomb marking *stelae*[4] (Bard 2000, p. 70; Wilkinson 2001, pp. 80-81), sarcophagi and statues (Stocks 2003). The use of basalt for both elite equipment and potter's wheel bearings could signify wider changes within the fabric of Egyptian society, beyond the creating of pottery, such as who was determining the use of the potter's wheel in the first place and why it came to be invented or introduced at all. The use of the potter's wheel could have represented a form of control by newly established elite classes, perhaps demonstrating their power and perhaps dominion over others. It could perhaps signify close technological links to foreign nations such as Canaan, Palestine and Mesopotamia, and such links between these ancient societies will be examined in Chapter 4.

Chapter 5 will investigate how the potter's wheel might have come to Egypt. It is commonly assumed that the potter's wheel was not invented in Egypt but in the Near East (Kuhrt 1995, p. 22; Freestone and Gaimster 1997 p. 15). Consequently, this chapter will assess if this was the case and if so, why. Through examination of technological and economic theory and the uptake of innovations such as the potter's wheel, it is hoped to better understand why the Egyptians introduced the potter's wheel at all. Arguably, the Egyptians had been successful in creating far superior pottery by hand (even relatively coarse wares) for centuries before the introduction of the potter's wheel (e.g. coil-made Black topped Badarian wares of Naqada I-II A/B (Petrie & Quibell, 1896, pp. 12, pl xviii-xxi; Petrie & Mace, 1901, pp. 13, pl xiii; Sowada, 1999, pp. 85-6)). In contrast, the use of the potter's wheel usually denoted a deterioration in the decoration and beauty of the pottery in favour of rather plain, utilitarian-style pots (Freestone & Gaimster, 1997, p. 15).

Chapter 5 will try to make sense of this rather odd trend away from decoration and will investigate if there are other underlying political reasons for such a change in technology. It is proposed that the reason for the invention of the potter's wheel was not to mass-produce utilitarian wares, but rather to create specialised vessels made on a

---

[3] The *Satire of the Trades* claims the potter "is muddier with clay than swine to burn under his earth," *Sallier Papyrus* II, Column V, line 5 (Parkinson, 1999, pp. 273-83) e.g. BM10182.

[4] *Stelae* or *stele*, from Latin "*to stand*" is the term Egyptologists use to refer to an upright stone slab or pillar bearing an inscription or design and serving as a monument or marker.

specialist piece of machinery. Using selected case studies, it is proposed in Chapter 5, to consider the arguments for the mass-production of pottery vessels and ascertain where the first wheel thrown pottery was located. The changing traditions of styles and forms of shaping pottery will be studied with the view to determining the extent to which the potter had a choice in their methods of shaping pottery, or whether this was controlled by the elite state officials.

Chapter 6 will examine pottery of the early Old Kingdom (c.2686-2181 B.C.) to ascertain when the potter's wheel was in use, what pottery types the potters were creating with their wheels and in what contexts they occurred. Once possible wheel thrown pottery has been identified through examination of museum pieces, Chapter 6 will consider to what extent the use of the potter's wheel can be noted on pottery. Through practical experimentation by manufacturing replica pottery using a reconstructed potter's wheel based on pictorial, literary, ethnographic work and excavated potter's wheel bearings, as outlined in Chapters 2 and 3, it will be possible to deconstruct the manufacturing methods used by the Egyptians to create wheel thrown pottery. From these experiments, a greater understanding will be gained of how to determine what manufacturing processes were involved in the excavated pottery assemblages. A fresh perspective will therefore be achieved for analysing and examining wheel thrown pottery and a greater understanding as to why the potter's wheel was developed as an invention.

By undertaking experiments in understanding the techniques of throwing on the potter's wheel, the aim is to resolve the terminological problem of what constitutes a vessel thrown on a hand-spun potter's wheel when compared with a vessel that has been formed by coiling. The methodology employed for the experiments will involve firstly creating coil and wheel thrown pots, so as to enable the author to identify the macroscopic details indicative of manufacture. The resulting pots will be photographed and X-rayed to provide further insights of manufacture. The methods will be filmed and photographed in order to deconstruct the gestures and movements made during manufacture and ascertain whether the techniques used could be associated with particular manufacturing marks produced on the pots. This criterion of manufacturing marks would then be compared to archaeological pottery collections in museums to identify potentially wheel thrown pottery using the characteristics of wheel throwing and coil-building which had been identified in Experiment 1. Experiment 2 will then involve the replication of a known potter's wheel in the British Museum collection, employing it for throwing selected vessels and testing the results by comparing the macroscopic features.

Given that it is likely that the potter's wheel was instigated through elite sponsorship (as postulated in Chapter 5), in Chapter 7 the contextual evidence of the vessels will be assessed to establish how the potter's wheel was used to create pottery. If the potter's wheel was used to create vessels for the elites, it is likely that wheel thrown vessels would only occur in elite contexts, such as in ritual or funerary offerings. In Chapter 6, the pottery of the early Old Kingdom will be examined to ascertain when the potter's wheel was in use, what pottery types the potters were creating with their wheel, and in what contexts they occurred. Early wheel thrown vessels occurred in similar cultic and funerary contexts in Levant and Mesopotamia (Courty & Roux, 1995) and it appears that the Egyptians adopted this new technology to produce items in similar contexts (funerary and cultic) but in an Egyptian manner. Social and economic literature and technological theory relating to the uptake of this new technology will be assessed and the reasons behind the use of the potter's wheel analysed. The Egyptians seemed to utilise this new technology to produce their own version of miniature vessels previously made in stone. The traditional methods of hand-building pottery vessels were successful in producing pottery items of high quality on a large scale for the domestic market, so it would seem that the potter's wheel was a rather redundant invention. It is anticipated that by investigation of the location of pottery production, whether in an industrial workshop or domestic area, and by considering how it was being made (wheel or hand, or partially by hand and finished off on the wheel) and how it was being fired (open or so-called 'bonfire firing' or enclosed updraught kiln), that this will indicate whether the use of the wheel was inspired by elite sponsorship. The use of basalt for the potter's wheel bearings also appears to be significant, given that it was usually restricted to royal building materials and items such as statuary, temple floors and sarcophagi.

By examining theories of innovation, technology and technical systems in conjunction with ethnographic research and analysis of the manufacturing marks of selected Egyptian pots from various sites and sources, it is hoped to identify the origins and use of the potter's wheel in Egypt. It is conjectured that the potter's wheel was adopted from Mesopotamia and the Levant regions and this research will address when this occurred, attempt to understand how this transition took place, and consider the underlying processes and effects, to ascertain why these might be significant. Through analysis of manufacturing marks on pots, it is planned to deconstruct the various manufacturing techniques that the Egyptian potter had to learn and to replicate those in experimental reconstructions using replica potters' wheel bearings based on the Egyptian standard. Understanding the techniques that the Egyptian potter had to master, combined with the pictorial, textual and circumstantial evidence, it is anticipated that new insights into the production and organisation of ancient pottery workshops will be apparent.

Chapter 2:
Seeking the Potter's Wheel

As outlined in Chapter 1, it has yet to be determined exactly when the potter's wheel began to be used either in Egypt or the Near East. The potter's wheel is often thought to have originated in Mesopotamia in the 4th millennium B.C. and subsequently its use spread to the Levant and Egypt (Freestone & Gaimster, 1997, p. 15; Kuhrt, 1995, p. 22; Pollock, 1999, p. 5; Simpson, 1997a, pp. 50-5). The first use of the wheel was considered to be for specific cultic contexts since wheel-finished pots are regularly excavated in temple sites in the Near East (Roux, 2003, pp. 15-18; Roux & de Miroschedji, 2009, pp. 155-157). It is commonly assumed that the potter's wheel was utilised solely as a mechanism for creating standardised mass-produced utilitarian wares (Bourriau, Nicholson, & Rose, 2000, p. 142). However, this may not be the case in terms of the first usage of the potter's wheel, even if it was ultimately employed in mass-production. The initiation of such a technology often requires some sort of impetus from another source such as the royal courts (Papazian, 2005, p. 76) or temples (Janssen, 1975, p. 183) before it can be instigated. An improved chronological framework needs to be established in order to identify when the potter's wheel first began to be used in Ancient Egypt in comparison to the rest of the Near East and to enable further analysis to be undertaken. In order to understand the chronological significance of the first use of potter's wheel, a chronology of the Near Eastern and Egyptian Periods is illustrated at the beginning of the text.

Previous Literature Relating To the Origins of the Potter's Wheel

Before establishing a chronology for the first use of the potter's wheel, it is useful to ascertain how previous scholars have discussed the evidence for the origins of the potter's wheel. Despite at least one hundred years' work on the study of Ancient Egyptian ceramics by archaeologists, there is little research on the underlying manufacturing processes involved in the production of Egyptian pottery, nor on the origins of the potter's wheel in Egypt. The focus of research has often been on the decoration and function of the pot rather than on the manufacturing methods used. In the early part of the twentieth century, any discussion on the potter's wheel was often a brief comment indicating that the wheel seemed to be in use and ignoring how it came to be used as a technology. For example, Petrie stated, "The first use of the wheel regularly is for the great jars of the royal family in the first dynasty" (1925, p. 57). Reisner (1923; 1931, pp. 174-5) dated the use of the wheel in Egypt between the reigns of Khasekhemui (last king of second dynasty c.2650 B.C.[5]) and Sneferu (first king of fourth dynasty (c.2640-2604 B.C.)) although without any discussion as to why he thought this was so. Singer, Holymyard and Hall (1954) were uncertain as to whether the wheel originated from one centre in the Near East or several.

Frankfort (1924, p. 7) and Junker (1929, p. 125) positioned the earliest use of the wheel in Egypt during the reigns of the 4th dynasty kings of the Old Kingdom without explanation. Frankfort described the use of potter's wheels in contemporary Crete as a *tournette* or turntable, where a slowly spinning wheel disc is supported on a pivot and is used to rotate the vessel in order to build up coils rather than being used as a throwing device. Frankfort also described the use of a cart-wheel shaped wheel that is rotated with a stick by modern Hindu potters to build up sufficient momentum to throw several pots. Frankfort (1924) briefly mentioned some of the evidence for a potter's wheel being used during the Old Kingdom in Egypt. He proposed that the bases of some of the pottery vessels at this time were finished with a knife and suggested that the vessels were finished on the wheel to two thirds of their height, and then the lump of clay was cut off with either string or a knife. However, he thought that the potter's wheels of this time were the same as the Cretan examples he had discussed previously, and was of the opinion that the Egyptians continued to use the *tournette* until the time of the Ptolemies (c.323-30 B.C.). He postulated that in order to be a true potter's wheel, the wheels had to be attached to a flywheel operated by the feet and that this was never used by the Ancient Egyptians (Frankfort, 1924, p. 7).

One of the first archaeologists to discuss the potter's wheel in any great depth was V. Gordon Childe who in his 1954 treatise *Rotary Motion,* discussed the use of the potter's wheel as a technology utilising centrifugal force. He examined potter's wheels found in Crete, Mesopotamia, Israel and Greece, which consisted of wood and fired clay discs c.90cms in diameter (Childe, 1954, p. 201) and hypothesised on their function. Childe postulated that the use of forging metal to make a saw was fundamental for the construction of such wooden wheel-heads and considered that potter's wheels could only have coincided with the beginning of the Bronze Age, or more importantly the first use of copper to create tools (Childe, 1957, p. 3). Childe (1954, pp. 194-195) also discussed the important changes that can be detected upon pottery when the potter's wheel is used during manufacture. The potter's wheel leaves characteristic concentric ring marks on the pottery which can easily be detected, and is the result of supplying centrifugal force to a lump of plastic clay. He thought that 100 revolutions per minute would be required to achieve these rings. Childe (1954, p. 194) believed that

---
[5] Dates based on Oxford History of Ancient Egypt (Shaw, 2000) dates, whose author cautions that dates are often uncertain before 664 B.C.

these concentric ring marks did not explain much about the machine that made them. He also discussed the need for wheel bearings to allow the potter's wheel-head to spin freely, but thought that these would be made of wood, which would explain why no complete ancient examples had been found. As a result Childe did not recognise that the wheel bearings were in fact made of stone such as limestone or basalt, and comprise of a socket and pivot with a clay or wooden wheel-head placed on top, as identified by Powell (1995). This is perhaps the reason why so many are mislabelled in museums as door sockets (Egyptian Museum 72365), quernstones (Egyptian Museum, room 34 C 13.1248) and even olive presses (Brewer, Redford, & Redford, 1994, pp. 19, fig 4.10) see Table 2.2. Illustrations and photographs of selected potter's wheel bearings from museum collections are described in Table 2.2. The examples included in the Appendix are the models chosen by Powell (1995) and the author to be reconstructed for wheel-throwing experiments (BM 32621 see Chapter 6). Additional examples that were previously unpublished or mislabelled have also been included.

Childe (1954, p. 196) was also one of the first researchers to note the problems of translating *tournette* as "pottery disc" or wheel and *tour* or *tournage* as "potter's wheel bearings" or "slow wheel," both being distinct terms in French. In English these are somewhat confusing labels as both are capable of spinning sufficiently to centre the clay and could both be called "potter's wheels". Childe suggests that a more sensible suggestion would be to designate the *tournette* as a turntable (i.e. not utilising centrifugal force and where the pot is built rather than thrown) and *tour* as the potter's wheel (utilising centrifugal force and where the pot is thrown). Childe brilliantly sums up his irritation with these translation issues, "Unfortunately, English archaeological literature has been bedevilled by the translation of the French *tournette* by the self-contradictory term 'slow wheel'" (Childe 1954, pp. 196-7). Regrettably these French-English classifications still occasionally occur within the archaeological literature and continue to cause confusion (see Figure 2.1).

In 1959, Foster, following Franchet (1911), proposed an evolutionary sequence of the potter's wheel from the solid to a pivoted turntable, and from the simple to a double wheel. The simple wheel is considered to have been invented sometime in the 4th millennium B.C. in Asia and was associated with the adoption of the wheel throwing technique. Foster (1959a) suggested that the speed of the production of pottery was the key reason for its invention when using the wheel. It was assumed that the wheel would have made it possible to mass-produce standardised vessels often described in excavation reports of sites dating to 4th-3rd millennium (Foster 1959b, p. 101). This reason for the development of the use of wheel has been widely accepted by most anthropologists and archaeologists to be a continuous phenomenon (Edwards & Jacobs, 1987; Blackman, Stein, & Vandiver, 1993, pp. 63-7; van der Leeuw, 2002, pp. 238-288). However, this explanation was refuted by Courty and Roux (1995; Roux 1990; 2008; Roux & Courty 1997) who suggested that the first use of the wheel, at least in the Near East, was used for shaping rather than throwing vessels. The Near Eastern potters made coiled "roughout" vessels and then smoothed and finished the pots on a wheel (see Figure 2.1). They suggested that there is no evidence for the use of throwing

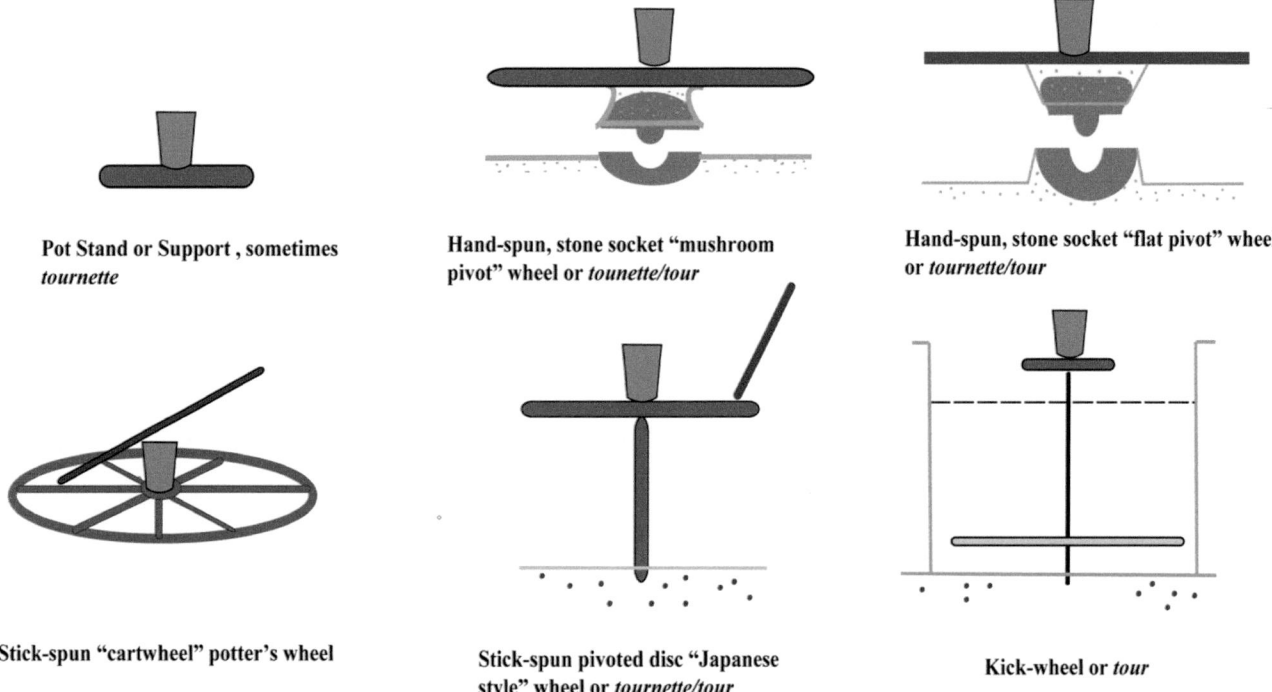

FIGURE 2.1: DIFFERENT TYPES OF POTTER'S WHEELS WITH FRENCH MISTRANSLATIONS IDENTIFIED BY CHILDE (1954) *TOUR, TOURNETTE,* THE STICK AND KICK WHEELS. AFTER MILLER, 2009, PG 114, FIG 4.5. DRAWING: S. DOHERTY

on the wheel in Mesopotamia during the 3rd millennium B.C. Instead, they postulated that this agreed with Foster's notions (1959b) of the evolution of the potter's wheel since it represented a logical step between using a support or turntable to draw up the sides of a vessel and simple wheel throwing (Roux and Courty 1998, p. 748).

Lucas, in the pottery section of his 1962 work *Ancient Egyptian Materials and Industries* suggested that some of the necks of Predynastic pots may have been shaped on a "slow wheel" which Childe (1954, p. 197) had previously described as a turntable. The pots built up by hand on a table, mat or on the ground, may have produced these traces of wheel marks. He suggested that the first form of the wheel, a small circular turntable rotated by hand on a vertical pivot or shaft but with only limited momentum, would have been a development of this process and was convinced that wheel made pottery never fully displaced handmade wares. Lucas (1962, p. 369) highlighted the importance of the 5th dynasty tomb of Ty in Saqqara (see Figure 2.2.), which has the earliest representation of the wheel in Egypt (Épron & Daumas, 1939).

FIGURE 2.2: TOMB OF TY SHOWING POTTER WITH POSSIBLY THE EARLIEST KNOWN REPRESENTATION OF A POTTER'S WHEEL IN EGYPT, STOREROOM, REGISTER 7 SAQQARA, EGYPT C.2450-2300 B.C. (ÉPRON & DAUMAS, 1939, P. PL 71)

Previous Experiments on the Potter's Wheel

In the 1960s, experimental work by Amiran and Shendov (1966, pp. 85-87; 1984, pp. 107-122) opened up a new avenue of research into potter's wheels in Israel and Palestine in the Byzantine Period. These authors reconstructed the wheel bearings and wheelhead of a Byzantine wheel, comprising an upper and lower set of stones joined together with the aid of an iron hook, and were able to throw pots. This wheel reconstruction is now in the Museum Haaretz, Tel Aviv, Israel (Amiran & Shendov, 1966, pp. 85-87;1984, pp. 107-122). In 1986, in the Department of Pottery Technology *Newsletter*, Edwards and Jacobs (1986; 1987) extending the work of Amiran and Shendov (1966; 1984) recorded their experiments with stone pottery wheel bearings from archaeological excavations in Palestine.[6] They added a 30cm wooden wheel-head to the wheel bearings using a flattened cake of clay and despite adding graphite and machine oil as lubricant, they could not rotate their wheel more than 1½ revolutions per hand spin. Consequently, they determined that centrifugal force was not being induced and throwing was not possible. When these experimenters used an assistant to spin the wheel they achieved speeds of 15-20 r.p.m. and found that they could form a pot. However, they considered that the wheel bearings would only have been suitable for the forming and smoothing of necks and rims of vessels. Even increasing the diameter of the wheel-head to 40cms was not deemed sufficient for centrifugal force to be in action. They noted that 50 r.p.m. would be a suitable speed for achieving centrifugal force, without indicating the source of this suggestion.

Edwards and Jacobs (1987, pp. 53-55) are amongst the first authors to try to trace the development of the wheel from (1) pots being built on mats and being intermittently rotated while coils of clay are added (*Prototournette*). (2) pots made by the coiling technique on a slowly rotating wheel which was also used for finishing, smoothing and trimming the vessel (*tournette*). They suggest that the stone wheel bearings proposed by Amiran and Shendov (1966) or else a fired clay disc rotated on a peg could be used as a *tournette*. (3) Pots made on a *tournette* or slowly (15-20 r.p.m.) but continuously, rotating wheel. Edwards and Jacobs also noted some of the characteristic marks found on such pots: clear rilling marks formed by heavy finger pressure on the interior of the base and walls and a spiral torsion twist. They stated that although this is using the basic throwing techniques of the later fast wheel or "*tournage*," their experiments showed that the friction was too great for centrifugal force to be the positive shaping force in pot making. They considered that it was only with the development of the fast wheel or "*tournage*" and the use of centrifugal force that vessels can be termed truly thrown (Edwards & Jacobs, 1987, p. 55).

Colin Hope (1981; 1982; 1987a;1987b) was amongst the first scholars to research Egyptian potter's wheels and to consider how the wheels were put together and used. In 1987a, Hope published an article entitled *Experiments in the Manufacture of Ancient Egyptian Pottery* that was based on his work on the pottery of the Dakhla Oasis. He experimented with manufacturing examples of

---
[6] The authors do not mention from where the pottery bearings they used in experiments came.

Meidum bowls[7] and bread moulds, which he postulated to be wheel-made, although others think that they were made initially on a mould or *patrix* and then later wheel thrown (Arnold, 1993, pp. 21-24; Vandiver & Lacovara, 1985). In the experiments the authors successfully used an electric wheel and modern potters' tools (cut off wires, rib, brushes and wire ribbon tools) to make their bowls and bread moulds. However Hope (1987a, p. 105) determined that caution must be exercised in the identification of wheel manufactured pottery as many factors in both the manufacturing and firing processes can highlight or obscure characteristic wheel-made marks i.e. concentric striations, s-shaped base (see Figure 2.3), torsion marks, and that a trained potter is required to clarify these physical features (see Chapter 6 for discussions of these terms).

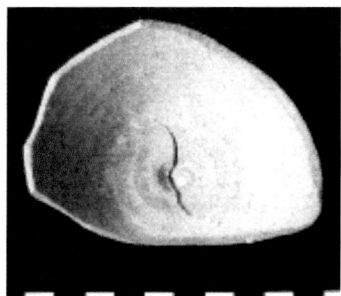

FIGURE 2.3: AN EXAMPLE OF AN S-SHAPED CRACK, INDICATIVE OF THROWN POTTERY, FROM GOBLET P03-219, TELL SABI ABYAD, SYRIA, LATE BRONZE AGE (DUISTERMAAT, 2008, PP. 379, FIG V27)

In 1995, a professionally trained potter, Catherine Powell, undertook experiments at the New Kingdom site of Tell el-Amarna as part of Barry Kemp's excavations. Powell published the Ancient Egyptian wheels in the British Museum and Ashmolean collections and, using the example of BM32621 (see Table 2.2), reconstructed the wheel bearings and attached an unbaked clay wheel-head. Using her wheel she successfully threw a variety of pots and bowls of New Kingdom types, achieving speeds of over 133 r.p.m. (far greater than that of Edwards and Jacobs (1987, p. 52) who only achieved 15-20 r.p.m. and who did not consider their pots to be thrown). The majority of the potter's wheel bearings in the museum collections of Cairo, Oxford and London comprise an upper pivot and a lower socket stone usually of basalt, granodiorite or limestone. They range from 15cm-24cm in diameter and vary in height from 5.5-6cm (Powell, 1995, pp. 309-311). Authors have suggested that a minimum of 50 r.p.m.[8] and maximum of 130 r.p.m. are the optimum speeds sufficient to throw pots (Amiran & Shenhav, 1984; Colbeck, 1982, p. 19; Rye, 1981, p. 74).

---

[7] Meidum bowls are described as carinated bowls with a bright red slip, polished with a round shoulder and rounded base. Commonly they are Nile B2 clay, see Vienna system in Appendix II and Chapter 7 for discussion and figures (Ballet, 1987, pp. 2-3; Op de Beeck, 2004; Wodzińska, 2009c, pp. 133-4).

[8] R.p.m. stands for rotations per minute.

Recent Scientific Research on the Potter's Wheel

To date, the most in depth work on the potter's wheel and indeed on Ancient Egyptian pottery in general is Arnold and Bourriau's seminal work *An Introduction to Ancient Egyptian Pottery* (1993). In Fascicle 1, Arnold has detailed the techniques of the manufacture of Egyptian pottery, and included a useful section on the potter's wheel termed "*central radial methods*". In this section Arnold (1993, pp. 36-83) has traced the available evidence for some sort of rotational device, as suggested by Bourriau (1981), that was in use on pottery from the Naqada I period (c.3600 B.C.) for the upper parts of the vessels (rims and shoulders) until the introduction of the kick wheel in the Roman period (c.30 B.C.). Arnold's analysis is thorough, she has examined tomb wall scenes of potter's workshops, limestone statues, wooden models and has identified some of the characteristics of wheel-made pottery and included useful examples. Arnold was not able to make a detailed study of Old Kingdom pots and was therefore unable to postulate when the potter's wheel was first introduced into Egypt. However, she places the first use of the wheel in Egypt at around the 5th dynasty (2400 B.C.) with increasing use during the 6th dynasty and later, citing the potter working at the wheel in Ty's tomb workshop as evidence (Arnold, 1993, p. 43). She has also suggested that pottery and stone production could have had early ties. The application of centrifugal force to the shaping of a stone vessel using Twist Reverse Twist Drills was probably in use as early as Naqada II (c.3500-3200 B.C.). The drill was weighted with netted stones bound to the shaft so that when the handle was turned the stones were flung around, driving more momentum than that which could be made by the craftman's hand alone, something confirmed by Denys Stocks's experiments when making such drills (2003, pp. 111-137, see Figure 2.4 and Chapter 4).

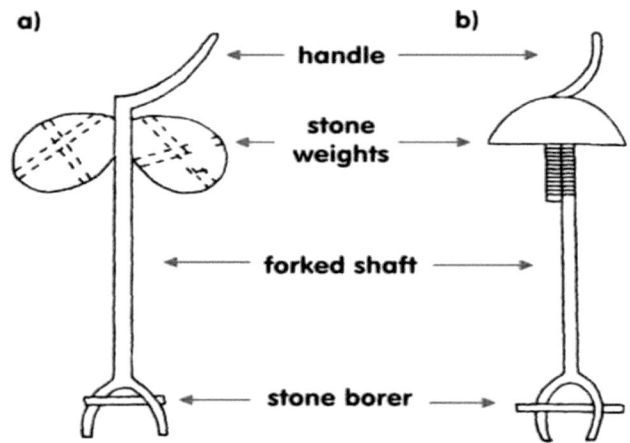

FIGURE 2.4: THE TWIST REVERSE TWIST DRILL. A) OLD KINGDOM EXAMPLE WITH 2 STONE WEIGHTS, GARDINER'S U25 DETERMINATIVE B) THE NEW KINGDOM VARIETY WITH A SINGLE LIMESTONE WEIGHT. THESE WOULD HAVE EITHER A FORKED SHAFT ATTACHMENT AS SHOWN, OR A HOLLOW BORER WITH COPPER TUBE ATTACHMENT. STOCKS 1993, P. 598

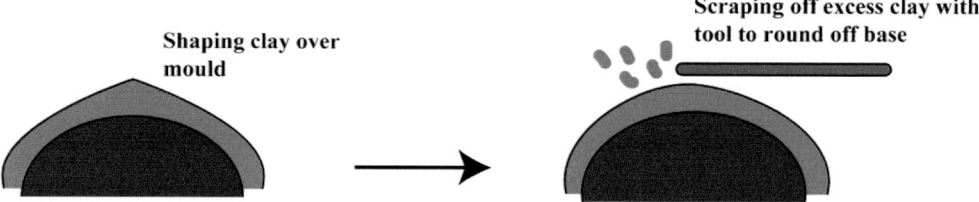

FIGURE 2.5: MANUFACTURING MEIDUM BOWLS OVER A MOULD OR FORMER. THE CLAY IS FIRMLY PRESSED OVER THE MOULD AND CURVED OVER. THEN A KNIFE OR SIMILAR TOOL IS USED TO SCRAPE AWAY EXCESS CLAY TO SHAPE THE BASE. DRAWING: S. DOHERTY

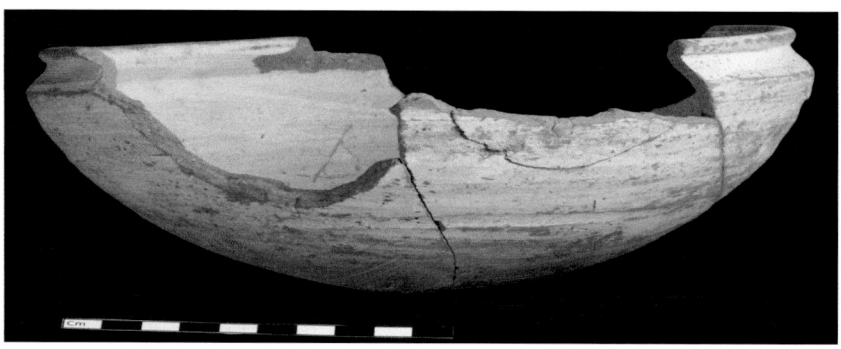

FIGURE 2.6: AN EXAMPLE OF A CD7 BOWL, 4TH DYNASTY, GIZA. AW9944-3. PHOTO A WODZIŃSKA

Scientific analysis has been performed on pottery in the form of xeroradiography[9] to detect manufacturing marks on pottery through images. Through such research Vandiver and Lacovara (1985, pp. 55-65) were able to establish a method for determining criteria for clay joins to the fractured edges of pots and pottery sherds (butt, bevels etc.). They also developed a classification system for the orientation of pores within the clay to determine the type of manufacturing and the sequence of construction. If the pores within the clay structure of the pot were in horizontal rows this indicated coiling, if there were a fairly even distribution of pores, this was a sign of throwing and if the pores were distributed around dicrete blocks, this indicated slab construction. Based on this research Vandiver and Lacovara were able to discover that Badarian black topped red ware vessels were constructed using sequential slabs. Several thousand pieces were analysed and the results indicated that none of the bodies of the vessels were made by coiling (though some bases were) and that Meidum ware bowls in early Old Kingdom were made by coiling (Vandiver and Lacovara studied an example from Giza of 5th dynasty date[10]). These bowls were initially formed in a mould and then later rotated on a support rather than on a wheel. Since the shape of the bottom of the bowl is round and could have been trimmed on a wheel, this suggested that the potter's wheel and a support mould were contemporary in Ancient Egypt. Arnold (1993, p. 21-22) agreed with this conclusion and suggested that Meidum bowls were probably formed over a hump or core similar to that used by modern potters in South America (see Figure 2.5). Vandiver and Lacovara suggested that complete throwing on the wheel occured during the Middle Kingdom 2025-1700 B.C. They based this conclusion on their study of a bowl of this date in the Museum of Fine Arts, Boston (MFA M173) since they detected three spiral throwing marks evident from the base to the rim of the vessel (Vandiver & Lacovara, 1985, p. 59). In contrast, some researchers e.g. David (2003, p. 324) have placed the introduction of the potter's wheel in the New Kingdom.

Anna Wodzisńka (2006) has also researched Meidum bowls and their possible connection with the potter's wheel in her analysis of CD7 bowls found during excavations at the tomb builders village at Giza (see Chapter 7 and Figure 2.6). These were produced on a vast scale in one location (the village) in a very short space of time, in the fourth dynasty (2600-2450 B.C.). CD7 bowls are a variant of Meidum bowls made of fine and medium fine Nile silt clay (NB2 in the Vienna system see Appendix II and III. These describe the differing clay fabric types of Nile silts and Marl clays as defined by the Vienna system, which is the recognised standard currently in use by archaeological ceramicists) and covered with a white wash. This is an unusual feature as Meidum bowls usually have a red slip applied before firing. These CD7 bowls seem to have had

---

[9] Xeroradiography involves the use of x-rays producing images on an electronically charged surface (Vandiver & Lacovara, 1985).
[10] Vandiver and Lacovara studied 37.2663, Museum of Fine Arts, Boston (1985, p. 59).

the very specific purpose of feeding the workforce of the 4th Dynasty Pharaohs' pyramid builders, independently confirming Frankfort (1924, p. 7) and Junker (1929, p. 125). Wodzińska has also found evidence that these CD7 bowls are unique amongst the pottery assemblage at Giza in that they were initially hand made and then later trimmed on the wheel. Many of these bowls show clear signs of being rotated on the wheel as there are concentric striation marks on many of the rims and shoulders of the vessels, and the bases are often irregular and trimmed (Wodzińska, 2006, pp. 405-429). See Chapter 7 for further discussion and analysis.

More recently, possibly as a reaction to xeroradiography studies, archaeologists and ceramicists have considered how the use of the wheel and the inherent manufacturing marks are reflected upon the pottery that they study. In her study of Old Kingdom pottery of the 6th dynasty, Rzeuska (2006a, pp. 35-54) devoted an entire chapter to pottery manufacture. She has included details such as the equal thickness of the walls of the vessel, noted that the temper added to the clay is usually parallel to the wall surface, and concluded "all the bowls, plates, stands and miniature vessels were made on the wheel," (Rzeuska, 2006a, p. 50). Therefore, by the 6th dynasty at least, the evidence demonstrates that the the potter's wheel is in common use for the production of funerary vessels.

Nonetheless, there is some evidence to suggest that the potter's wheel may have been in use at a much earlier date. Arnold (1993, pp. 41-9; Harpur 2001, p. 444; Holthoer 1977, pp. 6-26, and Odler in press) have discussed the various paleographic and iconographic sources for the potter's wheel. It is acknowledged that such secondary evidence must not necessarily be viewed as verification for the use of the potter's wheel for throwing pottery; only the manufacturing marks on the pottery can provide this. Consequently, given that there is insufficient evidence for the potter's wheel pictorially or in the physical remains of potter's wheels prior to the 5th dynasty, it is necessary to turn to the pottery itelf for more objective evidence regarding its use (see Chapter 6).

Archaeological Evidence for the Potter's Wheel

As has previously been noted, the literature presents a somewhat confused understanding of the use of the potter's wheel in relation to what the wheel looked like, where it originated and why it was introduced. It is evident that potter's wheels have been largely ignored by excavators and are currently variously labelled in museums as door sockets (e.g. Egyptian Museum 72365), quernstones (e.g. Egyptian Museum, room 34 C 13.1248) and even olive presses (Brewer, Redford, & Redford, 1994, pp. 19, fig 4.10 see Table 2.2). Sometimes potter's wheels are unpublished by the excavators (see Table 2.2), or not provenanced after being bought from an antiquities dealer e.g. the examples from the British Museum (see Tables 2.1 and 2.2. Descriptions of provenanced potter's wheels from Egypt and the Near East arranged chronologically are included in Table 2.1). In ceramic reports where workshop production and pottery specialisation are considered, even if manufacturing methods are discussed, the provenances of wheels are usually only mentioned in passing e.g. Dessel (2009, p. 124). Workshops and kilns are only occasionally found in archaeological contexts (see Appendix I, which includes a list of the most well-documented kilns, workshops and potter's wheels (where known) discovered in Egypt and provides chronological details and additional information of the excavated details of the site), and potting wheel bearings are even rarer (Tosi, 1984). Kilns and workshop areas could be located away from settlement sites in Egypt (McNicoll, Smith, & Hennessy, 1982, p. 57), such as at Hierakonpolis (Baba, 2006, p. 18; Hoffman, 1982) or the Dakhleh Oasis (Hope, 1979). The offsite location of such kilns and workshops would therefore hinder the likelihood of their being discovered by archaeologists (see Appendix I). Although, kilns occasionally occur near settlements e.g. Amarna (Nicholson, 1995b; 2010) and Medinet el-Gurob (Boatright & Hodgkinson, 2010; Hodgkinson, 2012; Shaw, 2011, p. 463), Shaw (2004, p. 16) noted that industrial workshops may not have been buildings at all and that many craft activities would have taken place in open areas or courtyards. Nevertheless, for pottery production it would be expected to find areas where some sort of roof or covering was supplied for clay storage and to keep it damp (as has possibly been located at Amarna (Nicholson, 1992) and Gurob (Hodgkinson, 2012, pp. 11-14). In lieu of caves, which have been documented as suitable sites for ancient pottery workshops in the Levant (Magrill & Middleton, 1997, pp. 68-73 and see Chapter 5) a roof would have been necessary for pottery production.

The next section of the present study is focussed on tracing the first examples of provenanced excavated wheels to determine whether the location of these wheels could provide insights into the pottery industry of ancient times. Two distinct types of wheel bearings seem to exist, the earliest examples are comprised of two bearings made of basalt, and pierced through, presumably so that they could be fastened to the ground using a wooden dowel. The later examples have the innovation of a socket and a pivot formed by the stonemason, removing the need for the wooden dowel. Instead of a "slow wheel," Childe (1954, p. 197) has suggested the term "simple wheel" to indicate a centrally pivoted disc of wood, stone or clay on a wooden frame. Of these, there were two varieties, the socketed and the pivoted disc. In the socketed variety, used in Crete, the pivot turns in a fixed stone socket in the ground and differs from the pivoted disc variety, used in Japan, in that the pivot is elongated to become an axle and requires an additional bearing above the socket. These two types will be discussed in greater detail below.

Pierced Wheelbearings

One of the earliest wheels so far discovered has been dated to the Chalcolithic period (4000 B.C.) at site 101 cave site at Tel Halif in Israel (see Table 2.1). Unfortunately, it is only

Site	Date	Location	Type of wheel	Material	Details	Picture	Reference
Tel Halif, Israel	Chalcolithic c.4000 BC	Site 101, Cave shelters	Pierced upper disk no. 2165, Lower basalt disk no. 2146. L100078, phase 10C 13.2cm dia, 5.6cm thick	Limestone upper and basalt lower	Under limestone slabs due to roof collapse. Assemblages included pottery, flint blades, stone grinding tools, beads and bone tools. Disk found next to several unfired pottery vessels and levigated green clay deposits.		(Dessel, 2009, pp. 20-22, fig 7; Jacobs & Borowski, 1993)
Meser, Palestine	Early Bronze Age IA c.3400 BC	Stratum I	Pierced flat Disk with biconical hole in centre c16cm dia x 6cm thick	Lower part	Upper face of disk is smoothed. Roughly v-shaped bowls also occurred within this stratum.		(Dothan, 1959, pp. 28, pl 2 F, fig 8:16)
Ur, Iraq	Early Bronze Age I c.3000 BC	Potter's quarter, east slope of hill	Wheelhead/Disk, 44kg, dia 75cm, 5cm thick	Fired (?) clay	Excavated at potter's quarter. Interpreted as probably rotated on a pivot using a stick, but could also fit onto domed pivot bearing. Quarter contained circular kilns with shallow fire pits 35cm deep x 90cm across supporting clay grates 1.3m in diameter.		(Simpson, 1997b, pp. 50, fig 1)
Megiddo Israel	Early Bronze Age I c.3000 BC	Megiddo stage IV, Stratum XVIII, locus , BB, prov 4014 Possible other clay wheelhead stratum XIII, in square O.14 and squares N 13 and E=T.3118. A fourth example comes from Stratum XIII A, prov 5058	Pierced Disk 19 x 5cm, and socket and pivot	Basalt and clay	Within area BB was a great city wall, associated temple buildings, location of potter's wheel within prov. 4014. Possible other clay wheelheads are buff coloured, some with red decoration and well burnished.		(Engberg & Shipton, 1934, p. 40; Loud, 1948, pp. 268, fig 13, pl 268:1; Wood, 1990, pp. 99, fig 1:1)
Jericho	Early Bronze Age	Square H. xiii-xiv	Basalt dia 25, T 3.7	basalt	Ex number 2904. Carefully shaped vesicular basalt. Symmetrical about the central hollow. Surface smooth and polished.		(Kenyon & Holland, 1983, pp. 560, fig 231 (2), pl 21)

TABLE 2.1: TABLE OF PROVENANCED POTTER'S WHEEL BEARINGS AND WHEELHEADS IN EGYPT AND THE NEAR EAST

Site	Date	Location	Type of wheel	Material	Details	Picture	Reference
TTel Dalit, Israel	Early Bronze Age I c.3000 BC	Area B, Stratum 2b. Locus 204Reg. 2182/1. Found in the "broadroom"	Pierced disk 1.1kg, dia 13-14cm, thick 2-4cm	Basalt	Disk reverse is dome-shaped, covered with chalk like incrustation (see fig a). Hole is funnel shaped from dome reverse and gradually narrows to 3cm. Glassy wear in one polished band on obverse (see fig b), 1cm wide, flat and smooth.		(Gophna, 1996, pp. 112-113, 144-5; Pelta, 1996, pp. 171-185, fig 1 & 2)
Tell Arad, S. Israel (Negev)	Early Bronze Age (3000-2650)	Stratum III, field no. 3961/53, locus 1555, level 50.10 (no. 12) and Stratum III field no. 857/51, locus 1085 level 51.91	c10cms in diameter, quite damaged	Basalt (no. 12) and chalk (no. 13)	7 circular stone objects hesitatingly named as wheels in the excavation report, but no. 12 and 13 pierced through, have lustre marks, no. 12 has traces of turning action		(Amiran, 1978, pp. 57, photo 123: 12-13; pl 77:12-13)
Beth Yerah (Khirbet el-Kerak), Israel	Early Bronze Age III	Beth Yerah II levels. (possibly SA Period D local phase 2)		Basalt?	Found in a level with mudbrick buildings. Associated pottery finds included band slip (grain wash) pottery, holemouth jars with ledge handles made of a gritty clay.		(Maisler, Stekelis, & Avi-Yonah, 1952, p. 170; Paz, 2006, pp. 95-100)
Tel Yarmuth Israel	Bronze Age III c.2600-2350 BC	Hypostyle Hall Palace B1, Area Bh, Square U39, Locus 1965	2 disks, upper disk (inv G-75/97, C.11805-1) 36.8cm dia, 4.8cm thick, lower disk (inv G-75/97, C.11805-2) 17.4cm dia, 3.8cm thick, Biconical perforation 3.2cmx2.4cm	basalt	Upper face is flat and polished, crossed by concentric polished rings whose width decreases towards outer rim of disk. These correspond to surface features observed on projection of the lower face of the upper disk. Lower face is rough with removal traces.		(Roux & de Miroschedji, Revisiting the History of the Potter's Wheel in the Southern Levant, 2009, pp. 158, fig 3)

The Origins and Use of the Potter's Wheel in Ancient Egypt

Site	Date	Location	Type of wheel	Material	Details	Picture	Reference
Tel Yarmuth, Israel	Bronze Age III c.2600-2350 BC	Area Ja, square K 41 locus 2104, precedes palace B1 and B2	2 disks, upper disk (inv G-65/99 C.16383-1) 26.5cm x 3.3. Lower (inv G-65/99 C.16383-2) 18.4cm x 4. Biconical perforation 1.8 x 3.2-4 at mouth.	basalt	Upper stone polished, traces of use wear. Lower disk is slightly convex and presents socket in form of a cone.		(Roux & de Miroschedji, 2009, pp. 161, fig 5)
Abu Sir, Egypt	Reign of Pharaoh Unas-Pepi II (c.2450-2181 BC) Early Bronze Age III	pyramid mortuary temple of Queen Khentkaus II, wife of King Neferirkare (c.2450-2300 BC) at Abu Sir. Associated with Pepi II cult	Wheelhead excavation no. 293/A/78,	Burnt clay, 45cm in diameter. Cracked and repaired.	Potter's workshop and kiln in the mortuary temple of Khentkaus II. Workshop had working table and potter's wheelhead. Wheelhead was found in secondary position on a ruin of low mudbrick wall on which was originally a slab of wood. Workshop surrounded by reed mats. Small storeroom for clay. Close to the door was a large storage jar with clay and broken mudbricks inside. Kiln found at opposite end of complex, next to the cult pyramid.		(Odler, in press; Verner, 1992; 1995, pp. 27, fig 27a, pl 5)
Nag Baba, Sudan	Middle Kingdom Early Bronze Age IV	Site 228. In Room VII had a pivot (73) near the east wall, either a door socket or a potter's wheel. It was lubricated with black resin.	Possible socket	unknown	Screen kiln located to south of workshop. Room IV had a thick layer of dung and straw, Room VIII had 47cm pit, 27cm deep containing mixed clay, walled with clay.		(Säve-Söderbergh, 1963, p. 58)
Jericho	Early Bronze Age levels	Possibly from Jericho city C.	Complete set bearings in basalt.		Childe only published the drawings of example and attributed it to Jericho. Its seems that Garstang (1934, pl. 19.2) mentions it coming from Jericho city C.		(Childe, 1954, pp. 201, fig 124)

Site	Date	Location	Type of wheel	Material	Details	Picture	Reference
Valley of the Queens	Middle Kingdom/ Late Bronze Age?	Tod?		Limestone			(Pers com Lecuyot July 2011)
Hazor	Late Bronze Age IIA (c.1400-1300 BC)	Area C, workshop part of set of shrine and cultic buildings. complete set of bearings in building 6225 (a), a second pivot was found in a storage room 6217 (b) and a third in room 6063 (c)	Set of pivot and sockets (ex no. C 1200/2, locus 6225 Stratum IB LB II) (a), and 2 separate pivots (ex no. C1201/2, locus 6217) (b) and (ex C1200/2 in locus 6225) (c).	Basalt	The potter's wheel bearings (a) were found on top of a platform or workbench c1.5 x 1m and 40cm high made of field stones, next to this was a pottery cult mask and pottery forming tools. The Upper pivot bearing (b) found in a storage room next to five broken pithoi. The second pivot bearing (c) was found next to a bench in a room that contained a cobbled floor area and a basalt bowl with pestle. The 3 sets of wheel bearings were all found within a larger potter's quarter, with open-fronted booths on the streets perhaps for selling their vessels. The workshops were located close to a stelae shrine and the Hazor city ramparts.		(Wood, 1990, pp. 16, 99, fig 1:8; Yadin, 1958; 1960)
Amarna	18th dynasty reign of Akenaton-Tutankha-men (1351-1323 BC)	Pottery workshop within industrial area Q48.4, area 12 "northern workshop" Eastern edge of main city of Amarna.	Upper pivot (3036) found within brick-lined pit/bin in Area 12, together with lumps of clay and sherds. A large zir filled with bricks had been sunk into the floor TA 87 Q48.4	Basalt 14.4cm dia, 2.4cm depth	Part of an industrial area within a rectangular enclosure associated with pottery production. Various kilns, puddling pits, clay storage areas and ash deposits. It may have been the supply centre to the workmen's village		(Nicholson 1992, p. 63; Rose, 1989, pp. 85-87, figs 4.2-4.4)

The Origins and Use of the Potter's Wheel in Ancient Egypt

Site	Date	Location	Type of wheel	Material	Details	Picture	Reference
Amarna	18th dynasty reign of Akenaton (1351-1334 BC)	Northern Suburb of Amarna, in largest house of the area T.36.11	Complete set of bearings ex. no.29/275, now in Ashmolean Museum 1929. 417. Pivot (a) 14.5cm diameter x 3.6cm (height); tenon 5.5 (d) x 1.5cm (h) 248g. Socket (b) 16.5 (dia) x 6.5 cm (h). Well of socket 5cm (dia) x 2cm deep 936.2g	Grano-diorite	The North suburb at Amarna contained a variety differently sized houses. It was not thought to have been an industrial area. House T36.11 is part of a series of similarly laid out large houses and probably was designated for an elite family and their household.		(Frankfort & Pendlebury, 1933, pp. 25, fig 6). Photos: S. Doherty
Lachish, (Tell ed-Duweir) Israel	Late Bronze Age III (c1200-1150 BC)	Cave 4034 in Grid square R 4. Upward course of the Wadi Ghafr towards Hebron, NE corner of entrance of tell, some distance away from the main public areas of the city.	Basalt pivot (tenon) field no. 6995 (a, b:12) and limestone pivot field no. 6994 (PM 39.834) (b: 13)	Basalt and local mizzi limestone	Large cave containing red and yellow ochre, lots of unfired sherd, heaps of prepared clay, crushed shell, charcoal, waterjar, mould for figurines. Potters' tools: bone points, pebble & shell polishers, sherd smoothed to use as ribs or turning tools. In Pit A contained the two pivots, upper surface of each is highly polished. Workshop contained a lower pit (B) which was reached via a flight of rock-cut steps was used for storing 40 fired vessels similar to those found at the Fosse temple and Structure III in the city. Later, the workshop was adapted into a sheep pen (layer 5)		(Magrill & Middleton, 1997, pp. 68-9,72, fig 6a; Tuffnell, 1958, pp. 291-3, pl 49:12-13)
Tell Sabi Abyad, Syria	Late Bronze Age	Pottery workshop comes from Level 6, squares N10-N13 to O10-13. Located within the settlement	pivot	basalt			(Duistermaat, 2008, pp. 349-353, fig V.7)

briefly mentioned by the excavators despite being labelled "of special interest" (Jacobs & Borowski, 1993, p. 69). At some point in antiquity, the cave ceiling had collapsed, leaving the well preserved chalcolithic floor littered with fired and unfired pottery, flint blades, stone grinding tools, stone beads, bone tools and the set of potter's wheel bearings. Termed as a "*tournette*" by the excavators, the potter's wheel bearings are comprised of a pierced socket made of basalt and a pivot made of limestone. Similar "pierced" wheel bearings (see Figure 2.7 and Table 2.1) have been found throughout the Southern Levant and have been dated to the Chalcolithic-Early Bronze Age I in various locations, such as Meser (Dothan, 1959, pp. 28, fig 8:16), Megiddo Stratum XVIII, XVI (Loud, 1948, pp. 268, fig1 & 2), Beth Yerah Stratum II (Maisler, Stekelis, & Avi-Yonah, 1952, p. 170) and Tell Dalit (Gophna, 1996, pp. 112-113, 144-5; Pelta, 1996, pp. 171-185, fig 1 & 2).

It would appear that the two pierced wheel bearings were a signature of the region during the Chalcolithic and Early Bronze Age I. These sort of wheel bearings were probably designed to hold a wooden pole placed into a heavy, fixed socket in the ground to steady it and reduce oscillation, with a wooden or clay wheel-head placed on top attached to the upper bearing. They were probably set up in a similar fashion to modern Japanese stick wheels (see Figure 2.8).

Similar style wheelheads have been found in various Mediterranean cultures, although without wheel bearings and were in use from 1900 B.C., much later than the Israel-Palestinian wheels. Wheelheads made of clay or wood have been noted in cultures such as Cypriote (Crewe, 2007, p. 211) and Cretan, where the clay wheelheads discovered ranged between 4-40kg and probably did not require the ballast of the wheel bearings (Evely, 1988,

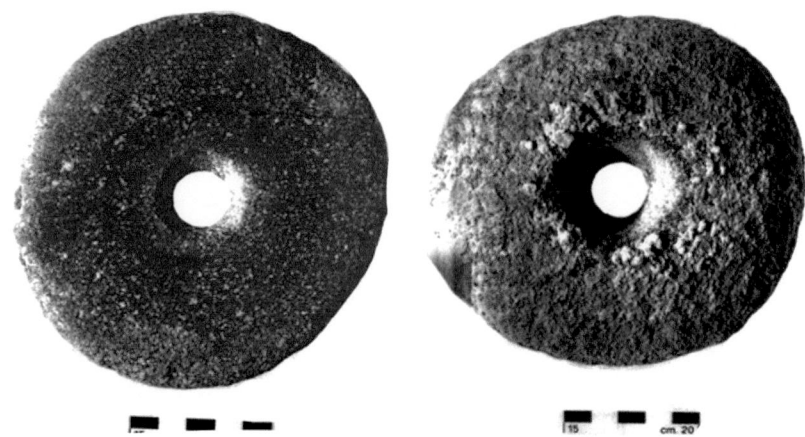

FIGURE 2.7: AN EXAMPLE OF THE PIERCED WHEEL BEARING FROM TEL DALIT, ISRAEL DATING TO THE EARLY BRONZE AGE I, AND (C.3000 B.C.) MADE OF BASALT. NOTE THE HOLE'S FUNNEL-LIKE SHAPE (GOPHNA, 1996, PP. 112-113, 144-5; PELTA, 1996, PP. 171-185, FIG 1 & 2)

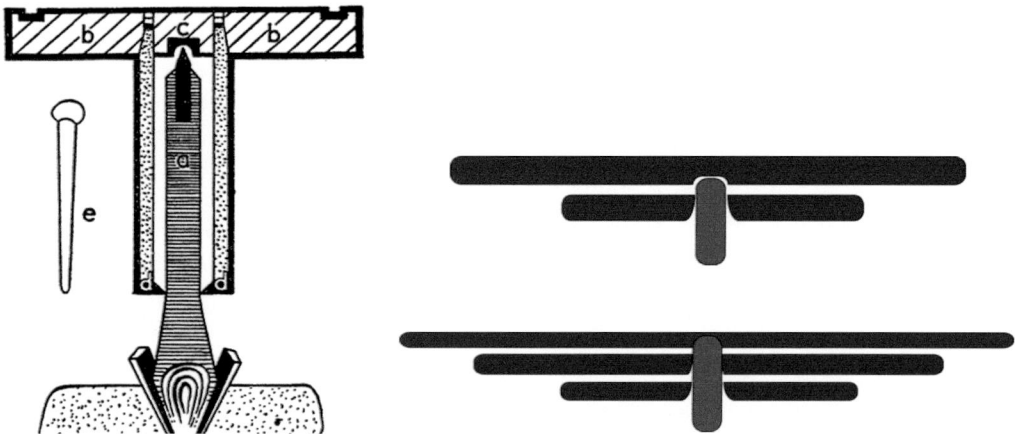

FIGURE 2.8: LEFT: JAPANESE POTTER'S WHEEL WITH SOCKETED DISK (A) HARDWOOD PIVOT; (B) HARDWOOD WHEEL WITH A PORCELAIN CUP AS BEARING; (C) HOLLOW CYLINDER EXTENDS DOWN THE PIVOT TO PROVIDE AN ANNUAL BEARING AT (D) WHICH STEADIES THE WHEEL; (E) STICK TO PLACE INTO THE NOTCHES IN THE WHEELHEAD TO SPIN IT. CHILDE (1954: 195, FIG 120). RIGHT: SUGGESTED ARRANGEMENT OF BASALT BEARINGS (LOWER DISK (S)), WITH A WOODEN WHEEL HEAD (UPPER DISK) ATTACHED, AND A WOODEN PIVOT. DRAWINGS: S. DOHERTY

FIGURE 2.9: CLAY WHEELHEAD FROM POTTERY QUARTER AT UR 44KG, DIA 75CM, 5CM THICK, C.3000 B.C. NOTE THE 8 PIN MARKS, POSSIBLY FOR AN ATTACHMENT TO THE SPINDLE OF THE PIVOT (SIMPSON, 1997B, PP. 50, FIG 1)

pp. 83-126; 2000, p. fig 11b; Xanthoudides, 1927); and the Cyladean, Lefkandi I and Tiryns cultures of Greece (Berg, 2007, p. 237; Wünsche, 1977, p. 27) and are very similar to the wheelheads found at Ur, which weighed up to 44kg (see Figure 2.9 & Table 2.1). However, unlike the Greek, Cretan and Cypriote wheel-heads, the wheelhead at Ur (see Figure 2.9) was likely to have been used with the contemporary pierced wheel bearings to support it. These pierced stone bearings would have given the wheel-head greater weight and stability and probably allowed the potter to spin the wheel at greater speeds for a sufficient time to throw a vessel as they were able to achieve a higher momentum.

Experiments with the pierced wheel bearing discovered at Tell Dalit (Pelta, 1996, pp. 171-185, see Table 2.1 for details and Figure 2.7 and Figure 2.8) suggested that it was likely to be the upper pivot section of the wheel as the hole in the centre is not of equal diameter, but tapered from the dome-shaped site (4-5cm) down two-thirds of the object's thickness (2cm) and then broadened again until it finished as 3cm on the flattened, glassy side of the bearing. Apart from a faint mark, there was no indication of rotation inside the bearing hole. As an experiment, a wettened wooden spindle was placed inside the hole of a replica of the bearing which, when dried, swelled up and wedged within the cavity (Pelta, 1996, pp. 177-9). Pelta, like other examiners of such pierced wheels, concluded that the example from Tell Dalit is a "*tournette*" or turntable (see Figure 2.1) as a forerunner to the so-called "Canaanite-Israelite fast wheel" which the authors suggested was introduced in the Middle Bronze Age (Amiran, 1963). It is likely that during Pelta's experiments she may have used too small a wheel-head[11] (the example from Ur in Figure

2.9 being 75cm in diameter) to achieve a sufficiently fast spin to enable throwing and therefore could only use the wheel as a turntable, and so did not achieve the wheel's full potential.

Another option for the design of the pierced wheels is derived from Tell el Yarmuth, in Israel c.2600-2350 B.C. (see Map at front of text) There, excavators Roux and de Miroschedjii (2009) found two complete sets of wheel bearings (see Table 2.1), with one of the wheel bearings being much larger than the other and which apparently side-stepped the need for a wheelhead. The wheel bearings were comprised of a set of large disc/pierced disc pivots made of basalt (one of c.27cm, the other c.38cm in diameter) and lower pierced discs c.17-18cm in diameter. These sets of wheel bearings were derived from the Palace B1 layers of the site designated as "*tournettes*" by the excavators (see Figure 2.1), and therefore were ideally suitable for fashioning and finishing the "Rotative Kinetic Energy" coil-built v-shaped pots (Roux & Courty, 1997; 2005; Roux, 2003). The wheel bearings were deemed not suitable for throwing, despite achieving speeds of 50 r.p.m. (Roux and de Miroschedjii 2009, p. 165).[12] These v-shaped pots and their significance to the use of The potter's wheel will be discussed in greater detail later.

Egyptian wheel-heads: Abusir

The first reliably provenanced wheelhead from an Egyptian context could be a contemporary of the potter's wheel bearings found at Tell el Yarmuth (Roux & de Miroschedji, 2009). Like the bearings at Tell el Yarmuth, the wheelhead uncovered during excavations at Abu Sir was found in a royal context, located in the mortuary pyramid temple of Queen Khentkaus II, wife of King Neferirkare (c.2450-2300 B.C.), and was likely to be associated with the cult of King Unas or Pepi II (c.2450-2181 B.C.).[13] A small pottery workshop was located to the north east of the pyramid temple with a kiln at the south eastern end, next to the *ka* pyramid (see Tables 2.1 and 2.2 and Appendix I). The kiln was conical shaped, originally 2m high and under 1m wide, with a firebox facing north. For a kiln the dimensions are small and since it does not appear to have traces of a perforated floor, it is unlikely to be an updraught kiln (see Chapter 5), but it does have traces of vitrified mudbrick. Around the kiln was found 5th dynasty pottery sherds and animal bones. A fragment of mud sealing dating to the reign of Unas was found in a nearby storeroom surrounded by kiln debris and malformed beer jars, suggesting a *post quem* date for the kiln area's initial construction, though the fill of the kiln itself contained only ash, sand and limestone chips. The kiln was propped up against the mortuary ka cult pyramid walls and is associated with a new entrance being made in the magazines opposite room SE-1 (Verner, 1995, pp. 33-4).

---

[11] Pelta (1996, pp. 179-185, fig 4-7) does not mention how large a wheel-head she used for her experiments, but based on the provided photographs, it was perhaps not much more than 20cms.

[12] Other authors have suggested that 50 r.p.m. is sufficient, (Jacobs & Borowski, 1993, pp. 53-55), but most cite Rye's 80-100 r.p.m. as a more suitable speed for throwing (Rye, 1981, p. 74).

[13] Another similar fired clay wheelhead has been uncovered in Sudan of Middle Kingdom date Stuart Tyson Smith *pers com.*

FIGURE 2.10: THE REMAINS OF THE POTTER'S WORKSHOP, WITH THE 20CM DEEP SLOT AND SHORT WALLS (A) AND (B).
AFTER VERNER (1995, PL 5, FIG 26). LABELS: S DOHERTY AFTER VERNER (1995, PG 26)

The wheelhead was found at the opposite end of the temple area, resting on a short wall MEW, one of two set against the enclosure wall MBW. Wall (a) was preserved to 85cm high and (b) to 56cm. Above wall (a) was a slot 20cm deep that had been cut into the wall MBW. The excavators suggested that this slot was used to insert a workbench which rested upon walls (a) and (b) (see Figure 2.10). Next to the Wall MEW was a shallow rectangular area measuring 3 x 1.5m, suggesting a fence line of palm ribs, or possibly a roof enclosed the workshop area. The wheelhead was made of baked clay, 45cm in diameter which had been broken and repaired in antiquity by drilling four holes near the broken edge and inserting string or wire (Verner, 1992; 1995, p. 26). In many ways it is not too dissimilar from the wheelhead discovered at Ur (see Figure 2.9) since it is made of clay and has a central depression for the pivot to be inserted into. However, it is smaller in diameter (45cm as opposed to the Ur Figure 2.9 example's 75cm). It is unknown how much the wheelhead weighed as it was excavated in the 1970s. It is now in the Egyptian Musuem in Cairo and there is only one photograph published by the excavators (Verner pers. comm. 2010 see Figure 2.11).

Odler (in press) recorded more details of the wheelhead. He has documented that it was red fired and there were traces of the black paint on the surface. The central hole (see Figure 2.11) seemed to be an original part of the object. Arnold (1993, p. 44) and Odler (in press) have suggested that a stick could have been inserted into this central hole to turn the wheel, but this would be very difficult. Most Japanese stick wheels are spun using sticks placed at the edge of the wheel in order to create sufficient momentum (see Figure 2.8). There were also traces of the plaster on the broken part of the wheel-head, possibly as an aid when the wheel was repaired by drilling holes (see Chapter 6 for further analysis).

As part of the general publication on the pyramid of Khentikaus, Verner has included a section on the pottery workshop and postulated on how the wheelhead might have been attached to the potter's wheel bearings.[14] Verner's (1995, pp. 27, fig 27b) interpretation of how the wheelhead was positioned on the pivot does not seem tenable as the pivot is not secure and is therefore liable to fall down (see Figure 2.11 right). The suggested socket (no scale) is too small to support the pivot and wheelhead. It is uncertain whether the pivot rotates or where the working facing is located, nor how the wheelhead is expected to stay on the pivot. There does not seem to be much evidence for Verner's interpretation, which is not discussed in the text. It is more likely that some form of wheel bearings were used to support the wheelhead. Unfortunately, no wheel bearings were uncovered during the excavations, so it is uncertain whether the bearings were of the pierced variety or the "late Bronze Age" tenon-pivot and socket variety. Figure 2. would suggest that the wheelhead central

---

[14] No potter's wheel bearings were discovered in Abu Sir.

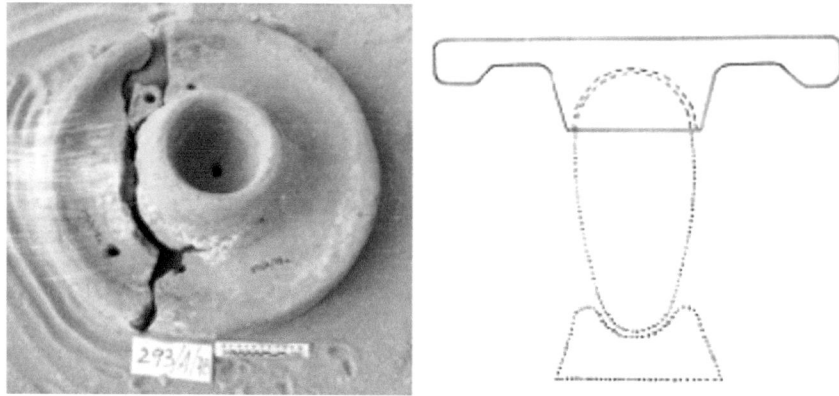

FIGURE 2.11: LEFT: THE ABU SIR CLAY WHEEL-HEAD. BURNT CLAY, 45CM IN DIAMETER. (ODLER, IN PRESS; VERNER, 1992; 1995, PP. 27, FIG 27A, PL 5) AND RIGHT: VERNER'S INTERPRETATION OF HOW THE WHEEL WAS SET UP (VERNER, 1992; VERNER, 1995, PP. 27, FIG 27B)

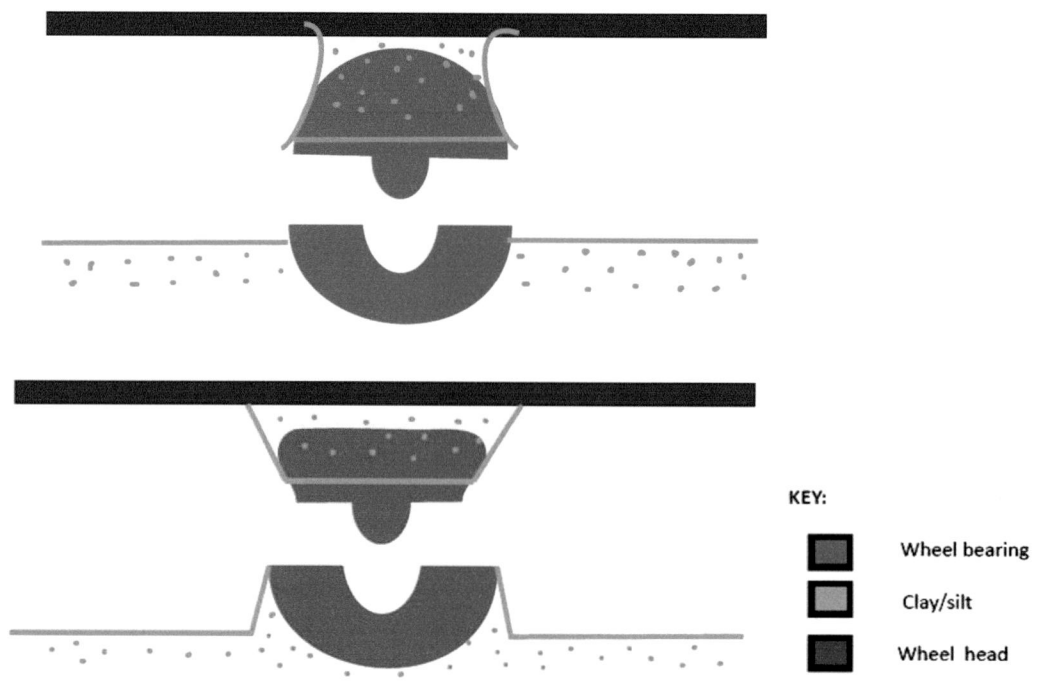

FIGURE 2.12: THE RECONSTRUCTED POTTER'S WHEEL, WITH THE PIVOT HEAD CURVED OR IN THE SHAPE OF A "MUSHROOM" IN THE UPPER EXAMPLE, FLAT IN THE LOWER EXAMPLES. DRAWING: S. DOHERTY

hole forms a curve, so it could perhaps be designed for a curved domed "mushroom" shaped pivot (see Figure 2.12 and table 2.2).

Tenon Pivot and Socket Wheel bearings

From the Late Bronze Age (c.1400 B.C.) a new form of wheel bearing seems to have become popular in provenanced pottery workshop sites throughout the Levant (e.g. Hazor (Yadin, 1958; 1960), Lachish (Magrill & Middleton, 1997, pp. 68-9,72, fig 6a; Tuffnell, 1958, pp. 291-3, pl 49:12-13), and in Egypt (e.g. Amarna (Frankfort & Pendlebury, 1933, pp. 25, fig 6; Nicholson, 1992, p. 63; Rose, 1989, pp. 85-7) and see Tables 2.1 and 2.2). Rather than two pierced discs of basalt, these comprise a lower "whirl" or rounded stone with a hollowed out "well" in the centre, shaped in a similar manner to a door socket and an upper domed (e.g. Figure 2.12 and 2.13) or flattened pivot stone with a raised knob or tenon shaped like a parabola (see Figure 2. and Table 2.2). This domed style pivot is perhaps what the wheel-head found at Abu Sir is designed for (see Figure 2.11), and possible evidence that these wheel bearings occured earlier in Egypt than previously considered). From the wear marks on the wheel bearings and Egyptian tomb decoration of potters (see next section), and compared to similar excavated wheels from

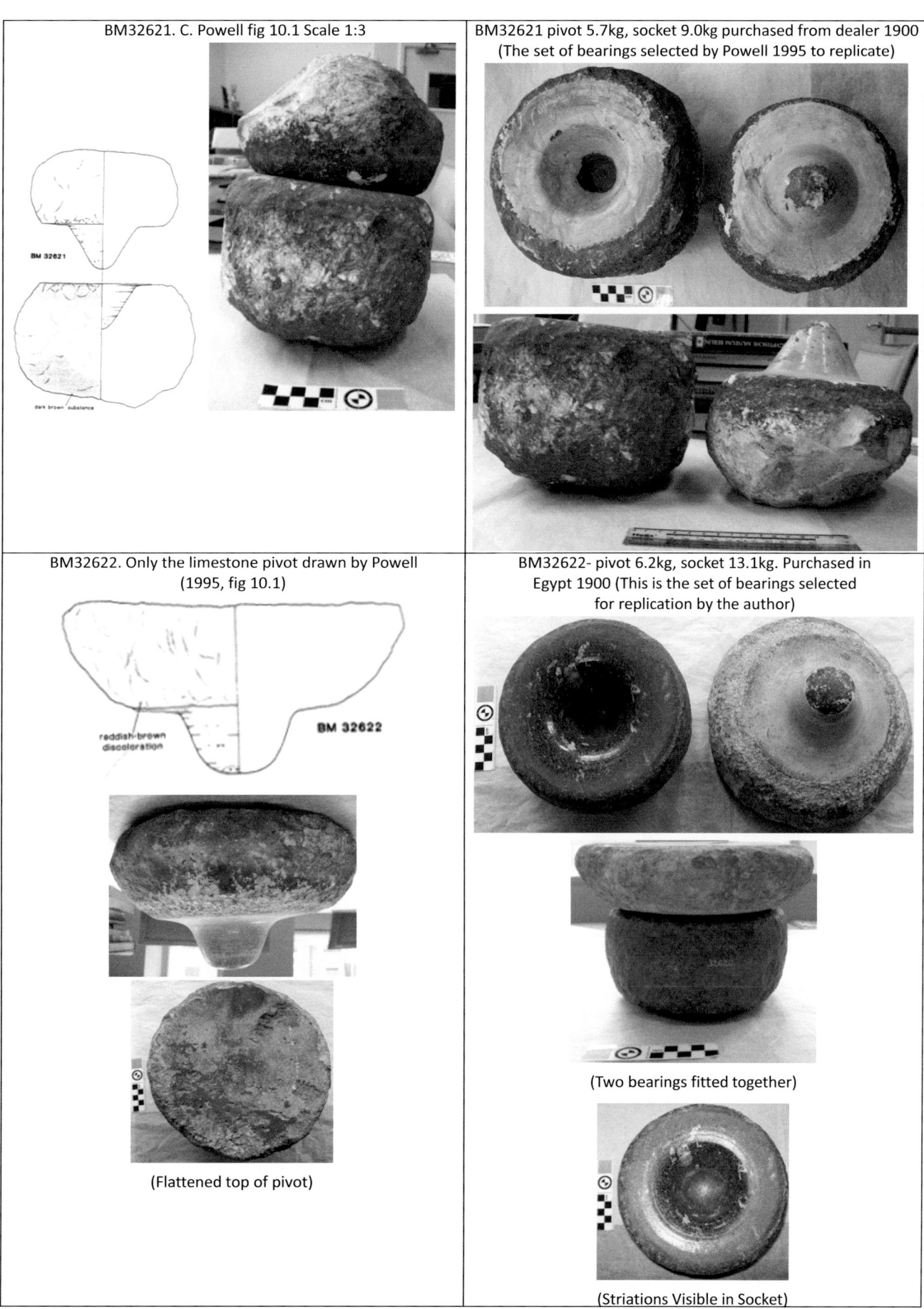

TABLE 2.2: SELECTED POTTER'S WHEELS IN WORLD MUSEUM COLLECTIONS AND PREVIOUSLY UNPUBLISHED EXAMPLES.
PHOTOS: S. DOHERTY UNLESS OTHERWISE INDICATED

The Origins and Use of the Potter's Wheel in Ancient Egypt

BM55316 C. Powell (1995) fig 10.1 3 part wheel bearing- limestone cap, pivot and socket 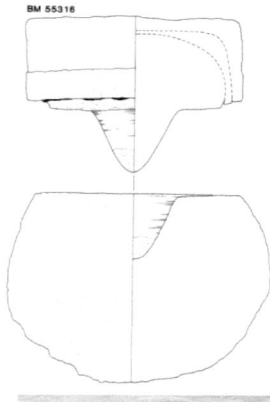  (the flat cap of the pivot as viewed on top of the socket, inserted over the pivot. Made of chalky material- comes off with a brush of the finger).	BM 55316 Pivot 5.8 kg, socket 13.1kg 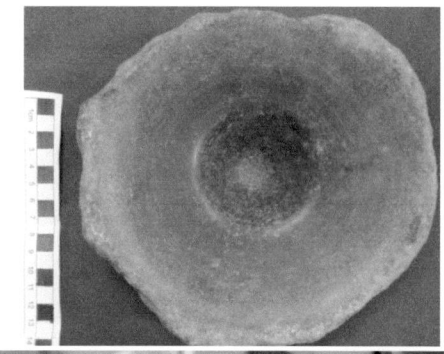 Socket  Pivot
Ashmolean collection 1929.419 Pivot 2.48kg, Socket 9.36kg (not recorded by Powell) Example excavated at Amarna by Pendlebury  Fitted together, note broken off edge of both socket and pivot, possibly indicating why it was discarded (?)	
Ny Carlsberg Glytotech AEIN 1186. Memphis, from Petrie's Excavations in Southeastern area, possibly Ptolemaic/ Roman but likely to be earlier Coarse outer surface, well used smooth inner surface. Base H. 6cm D14/16cm. Top H.6.7cm, D.12.5cm, Basalt or Granodiorite (Bagh, 2011, pp. 64, fig 1.76. Photos: Ole Haupt) 	

Potter's Wheel bearings described as a small olive press in the Agricultural Museum, Dokki, Cairo (Brewer, Redford, & Redford, 1994, fig 4,.10)	Unpublished example of limestone potter's wheel bearings from the Valley of the Queens. Courtesy of Guy Lecuyot

FIGURE 2.13: THE BRITISH MUSEUM COLLECTION OF UNPROVENANCED EGYPTIAN POTTER'S WHEEL BEARINGS. PHOTO: S. DOHERTY ©THE BRITISH MUSEUM

the Levant, it would appear that the pivot rotated on top of the socket. The raised tenon in the centre acted to keep the wheel running centred and true.

These two bearings (see examples in Table 2.2) formed a thrust bearing to effectively absorb the force parallel to the axis of revolution. Placing a baked clay or wooden wheelhead on top of the bearings added extra weight and increased the momentum of the spinning of the wheel. Pouring lubricant such as linseed oil (Powell, 1995, pp. 316, 322, 331-334) in the socket prevented the tenon from locking inside the socket and maintained an even spin. These wheel bearings have been called "fast" wheels by various authors e.g. Amiran & Shendov (1966), Kelso and Thorley (1943, p. 96), Johnston (1977, p. 206) and Wood (1990, p. 18) who have suggested that this type of wheel occurred from the Early-Middle Bronze Age (c.2000 B.C. i.e. the Egyptian Middle Kingdom). The title "fast" is meant to imply that the wheel bearings were spun by the hand of the potter fast enough to induce centrifugal force, which was apparently not "induced" prior to the Middle Bronze Age II (Wood, 1990, p. 18). However, as shown in Figure 2.1, many labels have been ascribed to these hand-spun wheels. The significance of centrifugal force and the techniques involved in throwing will be discussed later in Chapter 6.

The most recent and most complete publication of Egyptian potter's wheels is that by Catherine Powell (1995), in which she lists all the known wheels from the British,[15] Cairo and

---

[15] Although she does not fully publish BM32622- see Table 2.2, and she misses out the wheel found during the 1930s seasons Amarna in the Ashmolean collection, and that at Tell el-Daba a surface find of Manfred Bietak's excavations (Arnold, 1993, pp. 74, fig 87A).

Ashmolean Museums and the site of Amarna, together with their provenance (if known) their construction material and dimensions. Most of the provenances are unknown as they were bought from dealers in the 1900s (Marcel Marée *pers. comms.*). Some exceptions include the recent excavations at Amarna, where a pivot made of basalt was discovered in area 12 of grid square Q48.4 Phase I in a brick lined pit or bin (3036) in a workshop area containing puddling pit, clay, pot sherds and other potting paraphernalia (Nicholson 1992, pp. 62-63; Rose, 1989, pp. 82-86 and also Powell (1995, p. 310) for drawing publication). Another complete basalt potter's wheel was located by Frankfort and Pendlebury, (1933, pp. 24, 24 and pl XXX. 6) during the 1926-1932 seasons.[16] It was found in one of the most sumptuously appointed houses (T36.11) in the northern suburbs of Amarna, but its exact location within the house is uncertain (Frankfort & Pendlebury, 1933, p. 24). It is interesting to note that (provenanced) potter's wheels seem to occur in two contexts - either in a highly specialised workshop (e.g. Amarna at Q48.4 (Rose, 1989, pp. 82-86) or Lachish (Magrill & Middleton, 1997, p. 69) or within a large villa or palace-like structure (e.g. house T36. 11 in the northern suburb of Amarna (Frankfort & Pendlebury, 1933, p. 24) or palace B1 at Tell Yarmuth, Palestine (Roux & de Miroschedji, 2009, p. 157), and see Table 2.1. Some of the major problems with Egyptian wheels is that only a selected few have any provenance at all, most are recorded only in the most basic terms, and some are included as surface finds, so it is difficult to determine their context and age when no stratigraphic data is available.

Summary

Thus far in this thesis the archaeological evidence and the current state of the literature relating to the origins and use of the potter's wheel have been scrutinised, mislabelled examples have been identified and additional examples included. The provenance of these potter's wheels has been considered and an updated list is included in Table 2.2. In particular, through a review of the literature, the current thinking relating to the potter's wheel the evolution and development of the potter's wheel have been detailed as well as the reasons for its development. Several functional theories hold prominence for the use of the potter's wheel, namely, shaping and finishing of coiled pots, providing pots for the royalty and the elites, prestige in funerary and cultic contexts, standardisation of pot styles and mass-production.

As outlined in Figure 2.1, the problems of terminology used by scholars has been identified and there is evidence of confusion between the different terms used to descibe the potter's wheel. In particualr, there is specific terminology problems when the terms are translated, especially from French to English e.g. *tournette* as "pottery disc" or wheel and *tour* or *tournage* as "potter's wheel bearings" or "slow wheel," both being distinct terms in French. In English these are somewhat confusing labels as both are capable of achieving sufficient spin to centre the clay and therefore could both be called "potter's wheels".

The outcomes from practical reconstructions of wheel bearings have been examined and how these different wheels have performed when pottery of differing types have been made on them. Provenanced potter's wheels (as detailed in Table 2.1) have been described in term of material, dimensions, style and technical performance. In addition, the literature detailing the underlying manufacturing processes involved in throwing has been reviewed. Previous experiments in making and throwing pottery using replica (and actual) excavated examples from Egypt and the Levant have been discussed. These experiments seem to indicate that Near Eastern Archaeologists consider the potter's wheel would not have been utilised for throwing, whereas a professional potter working at Amarna, Powell (1995) suggests that the socket and pivot potter's wheels excavated at Amarna would have been capable of throwing. When these experiments are analysed in detail, the speeds being achieved by the potters would have induced centrifugal force i.e. between 80-150 r.p.m (Rye 1981) and therefore could be considered to achieve throwing. These differences in throwing capability could be explained by the design of the wheel bearings and this issue will be the subject of later Chapters 4 and 5. Two distinct types of wheel bearings exist. Where the provenance of wheel bearings are known, they seem to occur in elite contexts such as palaces or large estate buildings, particularly in the earliest examples. These details have been described in Table 2.1.

The literature relating to the marks characteristic of wheel thrown and coil-made pots have been considered. Provenanced wheels and the literature has been analaysed to search for the first use of potter's wheels. It has been established that by the 5th dynasty at least the potter's wheel is in common use for the production of funerary vessels. In contrast, the evidence is inconclusive for its use prior to the 5th dynasty. The importance of the 5th dynasty tomb of Ty in Saqqara is a significant piece of evidence. The literature reveals a variety of differing opinions regarding the date for the first use of the potter's wheel. The prevailing opinion seems to be that the invention of the potter's wheel could only have coincided with the beginnings of the Bronze Age, and the first use of working stone and copper/bronze tools. Potter's wheels were made from a range of different materials- baked clay, stones such as basalt or limestone which would have required different tools to work and procure the bearings. Thus far, the archaeological sources for the potter's wheel have been identified. Other sources of information such as texts, tomb decoration and funerary furniture have yet to be analysed. Such evidence could provide further details regarding the origins of the potter's wheel in Egypt and will be investigated in the next section.

---

[16] Excavation number 29/217, now registered in the Ashmolean collection as 1929.419.

Chapter 3:
Ancient Sources for the potter's wheel

Secondary Evidence for the Potter's Wheel in Egypt

The pottery workshop evidence has indicated that the potter's wheel "pivot and socketed" bearings were introduced sometime in the Early Bronze Age, however, some secondary evidence may suggest an earlier date. Therefore, by examining sources such as texts, tomb decoration, statues and wooden models it might be possible to gain a greater appreciation of the potting craft and further insight into when the potter's wheel might first have been used in Egypt.

Depictions of Potters at their Wheels: Tomb Scenes

Unlike the Levant and Mesopotamia, Egypt has a wealth of secondary evidence relating to craft as it was quite common for the elite members of Old Kingdom society to depict industry and craft activities relevant or useful to them during their lifetime which they would be desirous of having in the afterlife (Baines, 1994, pp. 71-90; Vasiljević, 2003, pp. 136-9). Such scenes should be viewed with caution since they are often embedded with multiple symbolic meanings, and should not always be read as simply being representative of "everyday" activities (Kamrin, 1999; Walsem, 2005, p. 69). However, some scenes go further and seem to represent accurate depictions of everyday life and could be used as a source of ethnographic information e.g. fishing and preparing fish (van Elsbergen, 1997; Nicholson & Doherty, 2014). The same could be said to be the case with pottery workshop scenes (Nicholson & Doherty, 2014). Potters working at their wheels and the vessels that they produced were occasionally depicted upon the walls of tomb owners and have been meticulously described by Holthoer (1977). During the Old Kingdom (c.2686-2181 B.C.), on the walls of a tomb, it was popular to describe common "everyday life" scenes that the deceased might have been associated with during their lives while not necessarily having been engaged in the activity personally (Gahlin, 2001; Vasiljević, 2003, pp. 136-7). The elite of the time became more of a person in their own right rather than just an extension of the Pharaoh's court. Previously, courtiers were often buried beside their king's tomb in secondary burials e.g. the 1st dynasty tomb of Aha at Abydos had c.30 subsidiary graves around it (O'Connor, 2011). However, in the Old Kingdom courtiers began to have their own tombs and to express themselves in the design and decoration within them.

One possible early example is in the Khephen's quarry, where the Giza tomb of Nebemakhet dating to the 4th dynasty (see Figure 3.1) depicts a potter working at his wheel and scraping off the excess clay with his hand.

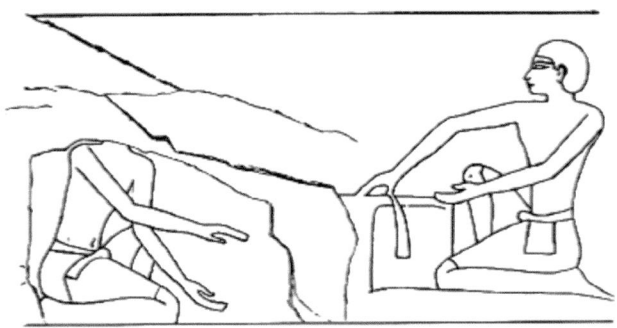

FIGURE 3.1: POSSIBLE POTTER'S WHEEL SCENE FROM THE ROCK CUT TOMB OF NEBEMAKHET, GIZA (AFTER HOLTHOER 1977, PG 6, FIG 1)

Unfortunately it is a badly damaged scene and no longer accessible so it is difficult to determine the particulars (Holthoer, 1977, pp. 6, fig 1). Do.Arnold (1993, p 43) suggests that it may be a leatherworking scene, but does not state her reasons. A second possible potter at his wheel comes from the early 5th dynasty mastaba of Kaaper (Bárta, 2001, pp. 166-168, fig 4.17). This scene is very faint, and has led Warden (2010, pp. 196-7, note 40) to suggest that this could in fact represent the pounding and grinding of wheat for flour.

During the 5-6th dynasties more credible and authenticated scenes of potters working at their wheels have been located, notably in the tombs of Ty at Saqqara (Épron & Daumas, 1939) and Ptahshepses at Abu Sir (Faltings & Vachala 1995, pp. 281-286; Vachala, 2004a; 2004b), in the 11th-12th dynasty nomarchs' tombs of Bakt III (BH 15, Dynasty 11), Amenemhat (BH2, Early Dynasty 12) and Khnumhotep (BH3, Mid-Dynasty 12) at Beni Hasan (Newberry 1893; 1894). Subsequently, a couple of examples have been located within the Second Intermediate Period of Horemkhawef at Hierakonpolis (Friedman, 2006, p. 20) and Kenamun at Thebes dating to the New Kingdom (Davies, 1930). The potter's workshops scenes are often located adjacent to baking and brewing scenes, suggesting to Dorman (2002, p. 58) that they ought to be viewed not merely as a pottery manufacturing scene in isolation, but as an important part of the food preparation and storage process; significant for the nourishment of daily life, but also for the maintenance of the deceased's *ka* in the afterlife (Drenkhahn, 1976, p. 87). These potters' workshop scenes provide a valuable insight into the everyday life of the Egyptian potter and suggest that potters were most likely to be attached to the great estates of the Egyptian nobility, rather than working alone. It is also possible that each village had its own potter who could create the pots that

FIGURE 3.2: POTTER'S WORKSHOP FROM TOMB OF TY, STOREROOM, REGISTER 7 SAQQARA, EGYPT C.2450-2300 B.C. (AFTER: ÉPRON & DAUMAS, 1939, P. PL 71)

the average Egyptian could not e.g. large water jars, and tableware, as is the case today in traditional potteries in Egypt (Nicholson, 2002; van der Leeuw, 2002; Vincentelli, 2003; van der Kooij & Wendrich, 2002).

The 5th dynasty nobleman, Ty, had a mastaba built at Saqqara during the reign of Niuserre.[17] Ty was Director of the Hairdressers of the Great House (i.e. the palace) and overseer of the estates and temples of Kings Sahure and Neferirkare (c.2440 B.C.). As such, he would have been involved with the day to day administration of the temples and estates and presumably organised the supply of pottery and its production, although probably indirectly (Épron & Daumas, 1939; Steindorff, 1913).

Above a scene of a bakery (see Chapter 5 and Figure 3.2) in the storeroom of Ty's tomb a potter's workshop is depicted with six potters busily manufacturing pots in two different ways, one of which was using the potter's wheel to make *hnw* vessels on the wheel (the bowls rather than the spouted vessel above the potter, see Figure 3.2), the others form a production line hand rotating pots and *dwiw* vessels (beer jars) in a stationary block. The hieroglyphic captions above the two potters making *dwiw* vessels reads ʿ*bb* "flattening, forming, smoothing, completing" and *ḳd* "building, forming"(Holthoer, 1977, p. 7). Both share a *dwiw* jar as their determinative, implying that this is what is being made (Hannig, 2003, pp. 265, 1343). Above is the potter at his wheel *dḥ hnw* "creating *hn* vessels." Therefore, in Ty's workshop we have representations of two pottery manufacturing traditions, namely, throwing on the wheel and handbuilding beer vessels using coils of clay and then rotating them in a stationary block to smooth down the joins and create the rim. Many beer jars of the Old Kingdom have pointed bases, testifying to the use of such a block as depicted in the tomb of Ty, rather than utilising a "turntable" or potter's wheel.

The kiln is placed to the far left of the scene, with bands around it to protect it from cracking when the mudbrick expanded during firing. A single potter supervises the kiln. He holds his right hand to his face as protection from the heat, in a similar manner to that common in bread making scenes (included on the lower register of the pottery scene in Ty's tomb). Above him is the caption *fš.t ṯ3*, "heating the oven/kiln." Ty's pottery workshop seems to provide evidence of specialised potters who were involved in the making of selected pottery shapes and that the potter's wheel was a significant part of that specialisation process (Costin, 1991; Longacre, 1999). In modern pottery production, potters specialise in particular shapes and often produce only a set number of vessel shapes, usually due to restrictions from market demand, despite being capable of more (Wodzińska, 2009a, p. 237). Nile silt clay potters in contrast to marl clay potters seem to produce a more varied corpus (Nicholson & Patterson, 1989 and see Appendix II and III).

The tomb of Ty has been dated to the end of the reign of Niuserre c.2450-2300 B.C. (Cherpion, 1989). The evidence from the tomb of Ty pushes back the date for the use of the potter's wheel in Egypt. Although in the Near East (as noted in Chapter 2) there is physical evidence for potter's wheel bearings in the form of pierced basalt disks, the earliest dating to the Chalcolithic period (4000 B.C.) at site 101 cave site at Tel Halif in Israel (see Chapter 2, Table 2.1); there is no evidence for these pierced disks in Egypt. As there are no known pierced stone wheels in Egypt, provenanced or otherwise, we cannot assume that the Egyptians ever utilised them. The earliest example of a clay potter's wheelhead in Egypt dates to the 5th dynasty of Old Kingdom, but provenanced pivot and socket wheel bearings are not known until the Middle Kingdom (see Appendix I). How then did the Egyptian artists know what the potter's wheel looked like? It may be that the shape of the wheel is derived from the artists' interpretation of another well-known and similarly shaped device, namely, the offering stand upon which provisions of food, perfumes and flowers were laid before the deceased e.g. Stela of master sculptor Shen (Faulkner, 1952, pp. 3-5, pl 1). Some 4th dynasty offering stands in the mastaba tombs at the Giza cemetery e.g. G1202 of

---

[17] Tomb no. 60 according to Jacques de Morgan or D22 by Mariette. It is located c 150m from entrance to Serapeum (Porter, Moss & Malek 2003).

FIGURE 3.3: STELE OF PRINCE WEPEMNEFRET BY NORMAN DE GARIS DAVIES
PHOTO: MFA BOSTON (SIMPSON & O'CONNOR, 2003, PP. PG 1, PL 1)

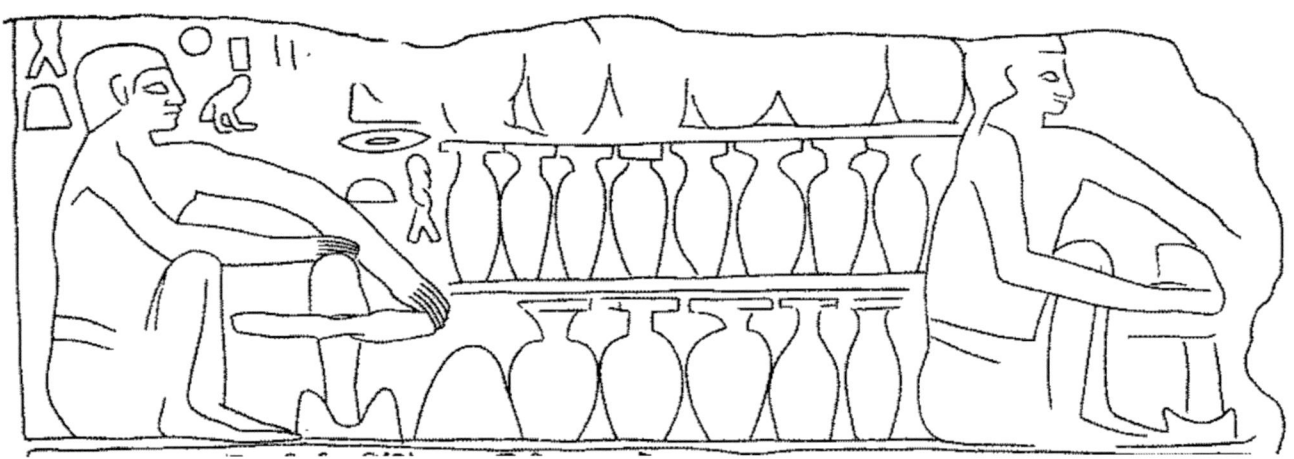

FIGURE 3.4: TOMB OF KHENTIKA FROM SAQQARA, IN THE CEMETERY OF PHAROAH TETI, DEPICTING
TWO POTTERS, 6TH DYNASTY. (AFTER: HOLTHOER 1977, PP. 8-9, HARPUR 2011, PP. 444-445)

Prince Wepemnefret[18] shows a clear colour demarkation between the red pottery stand and the white of the alabaster offering stand that it is placed upon (Fredrickson & Elsasser, 1972; see Figure 3.3). Other tombs such as the anonymous mastaba tomb G7650 (Flentye, 2007, pp. 292, fig 1) include depictions of offering stands very similar in shape to the potter's wheels depicted in the tombs of Nebemakhet, Ty, and Ptahshepses (Laurel Flentye pers. comm. July 2011).

In the 6th dynasty mastaba of Khentika from Saqqara, in the cemetery of Pharoah Teti (Holthoer 1977, pp. 8-9, Harpur 2011, pp. 444-445), a loose block depicting two potters was uncovered, though it may not originally be from this tomb, (James, 1953, pp. 34, pl XLII: XII [244])). This scene (see Figure 3.4) is used as an example of a simple low wheel (Arnold, 1993, p. 44) or the beginnings

[18] Museum cat no. 6-19825 Phoebe A. Hearst Museum of Anthropology, Berkeley, USA.

of a low pivoted wheel (Holthoer, 1977, pp. 8-9). The potter's wheels depicted in this scene do appear to be quite different from the one in the tomb of Ty (see Figure 3.2), with the socket for the wheel-head in clear view in the Khentika examples, but not in Ty. Do. Arnold (1993, p. 45) suggests that the pivot and wheel-head could have been made of wood which were slotted together with a tenon joint. She cites the example of the modern potters in Cyprus who use wooden pivoted wheels as an aid to their slab and pinch made vessels. The Cypriot examples have a short wooden axle with a wooden disc at the top and an iron point at the bottom which rotates in a socket of stone or metal, similar to the Japanese versions (see Chapter 2, Figure 2.8 and Childe 1954: 195, fig 120). The axle is held steady by a horizontal beam (Evely, 1988, pp. 83-126; 2000, p. fig 11b; Xanthoudides, 1927, pp. 123, pls 20b & 21). However, the potters of Cyprus never use their wheels for throwing, only to assist when building up coiled vessels. Sometimes the Cypriot potters rotate their wheels with their feet while forming the vessel with their hands, which is not attested in Egypt. The potters depicted on the loose block in the tomb of Khentika are not shaping or finishing a vessel, which one would expect had they been using a turntable, but instead they each have a lump of clay and are in the process of throwing a vessel. The potters have placed their right hands firmly on top of the clay and are spinning the wheelhead with their left, so that they can commence the centring process before the clay can be shaped into a vessel (see Chapter 6; video and Doherty: in press). While their potter's wheels might resemble the Cypriote turntables, they are being used in an entirely different manner. Although it resembles a pierced potter's wheel, it is likely that the wheel depicted in the scene of Khentika is of the pivoted and socketed type, but drawn in the style of an offering table.

Another loose block discovered during the excavations of the 5th dynasty Vizier Ptahshepses in Abusir depicts the relief of a potter squatting on the ground. The fragment (numbered I 204) was found outside the tomb (Vachala 2004ab, pp. 176, 179). The block was not part of the decoration of the mastaba of Ptahshepses, but has been dated to the 5th or 6th dynasty (see 3.5).

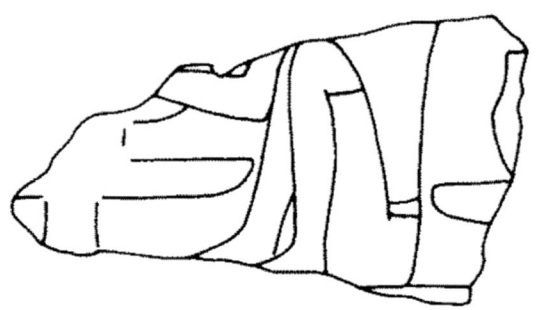

FIGURE 3.5: THE LOOSE BLOCK OF A SEATED POTTER WORKING ON HIS POTTER'S WHEEL, WITH A SECOND POTTER'S WHEEL WITH COMPLETED VESSEL TO THE RIGHT. AFTER: VACHALA 2004, P. 179, FRAGMENT I 204

There is rather a long time-gap before potters are again depicted on tomb walls (at least 120 years), and then the depictions only occur in two locales: in the tombs of the Nomarchs of the Oryx nome at Beni Hasan in Middle Egypt and in that of the Nomarch Djehutihotep at Deir el-Bersha (Newberry and Griffith 1895). Three tombs at Beni Hasan dating to the Middle Kingdom (c.2055-1700 B.C.), those of Bakt III (BH 15, Dynasty 11), Amenemhat (BH2, Early Dynasty 12) and Khnumhotep (BH3, Mid-Dynasty 12) each include detailed representations of potters and their workshops during this period. Consequently, Nicholson and Doherty (2014) have suggested that such scenes should be considered as ethnographic representations. Do. Arnold (1993, p. 46) proposed that the potter's wheels depicted in Beni Hasan represent a "newer" version, despite the potters having wheels very similar to one depicted in the tomb of Ty (see Figure 3.2), without the socket being visible, which happen to be depicted in the tomb of Khentika (see Figure 3.4). The 12th dynasty tomb of Djehutihotep has been celebrated as the first representation of the "tall stemmed wheel" where the potter sits on a chair in a similar manner to hieratic representations of potters (see below, *The written evidence*). This scene is similar to the so-called "birth scenes" or mammisi scenes of temples where the god Khnum is represented modelling the clay of the newly formed Pharaoh on a potter's wheel from the 18th Dynasty (c.1550-1085 B.C.) e.g. temple of Amenhotep III, Luxor (Brunner, 1964, pp. 68f, Pls. 6, 20).

The tomb of Bakt III (Newberry & Fraser, 1894, pp. 42-72, pl II, XXII-XXXVIII, see Figure 3.6) shows potting scenes in the fourth and fifth registers from the top of the western part of the south wall, in the main chamber beneath scenes showing, amongst other things, punishment of wrong-doers, procession of male and female dancers, men carrying funeral outfits of clothing, ornaments and weapons, stock-taking of asses, metal-smithing and games (Newberry & Fraser, 1894, p. 49; Porter & Moss, 2004, p. 153, nos 15-16 in the plan of tomb 15). The scene showing potters is exceptionally full and shows considerable detail (Holthoer, 1977, pp. 12, fig 14).

The pottery scene (see Figure 3.6) starts on the left with a wheel and goes on to clay trampling. Another scene shows kneading clay, carrying a cone or lump of clay toward the right where there follow two potters seated at their wheels facing right, another two with their wheels facing left and a further pair with wheels one facing right and the other left. In the lower register, a man is seen putting pots to dry, behind him and facing right is a damaged part of the scene though it is clear that it is of a man holding clay. In front of him is a man facing right and in front of his kiln whose fire can be seen through the stoke hole. To the right of that two more individuals are taking products from the kiln whilst another carries them away in baskets suspended from a yoke. Such a scene is all the more remarkable for its sheer quantity of potters working at their wheels, seven potters are represented here, all at differing stages of throwing pots.

The first potters at Bakt III's tomb are at their wheels sitting in a crouched position, with their knees drawn up against their bodies and with their legs either side of the wheel (see Figure 3.13).[19] The two potters at the end of the row are sitting opposite one another, with one knee raised, the other resting down, presumably for a sense of symmetry in an otherwise busy workshop scene.

Potter 1, to the far left, works at shaping and finishing the rim of his pot, potter 2 as suggested by Holthoer (1977, p. 12) appears to be lubricating the wheel since the potter is holding a grey substance (usually clay in these scenes, the same colour as the wheelhead) and is placing it in between the pivot and the socket. It is not certain that oil was used by Egyptian potters as a lubricant, nor is it certain whether

---
[19] This position is still common amongst Egyptian men today who can often be seen squatting down on their ankles.

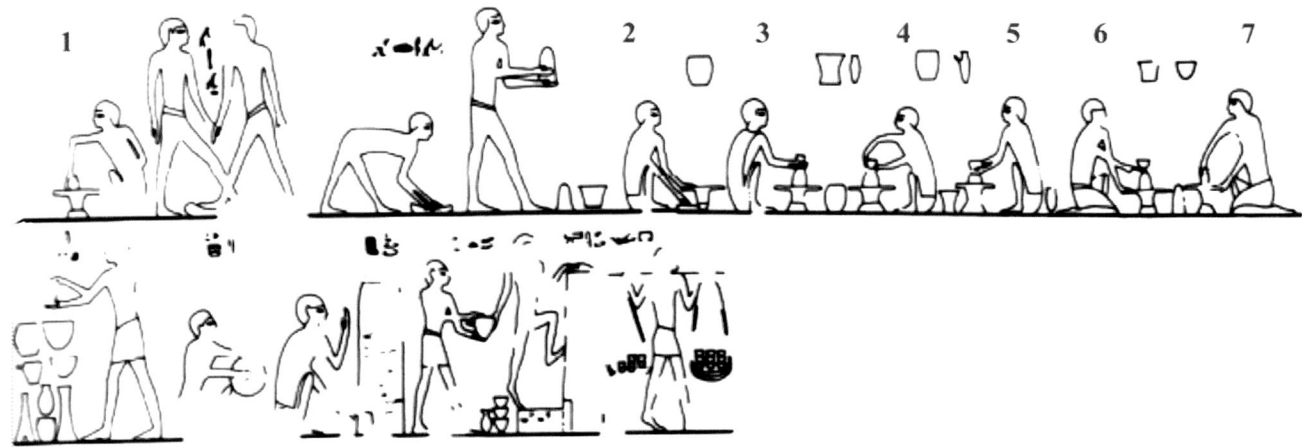

Figure 3.6: The tomb of Bakt III pottery making scene. After Holthoer 1977, pg 12, fig 14

lubricants were used at all. Some Egyptian wheel bearings e.g. BM 32622 and BM55316 have reddish brown discolouration on the working faces (Chapter 2, table 2.1), which perhaps indicates the use of a lubricant (see Powell 1995). If one follows Newberry's drawing (1893, pp. 43-50 pl VII), the potter appears to be adding coils of clay to his pivot, rather than lubricating the working faces. Holthoer (1977, p. 12) also differs slightly in his drawing with that of Newberry (1893, pp. 43-50 pl VII) and makes the pivot more obvious. If Holthoer is correct and potter 2 is indeed lubricating his pivot, then the grey substance may not be clay, but a lubricant. Diluted clay is not usually used as a lubricant as it would make the working faces of the wheel stick together as it dries. However, as the substance is painted grey in the scene, one wonders if this scene in fact depicts the addition of coils of clay to the pivot to secure the wheel-head before commencing throwing. Unfired clay is likely to crack in the heat if not kept damp and sometimes the wheelhead comes off if not properly secured. Some Indian potters add a layer of cow dung to prevent this occurring (Powell, 1995, p. 332).

Potter 3 at his wheel is commencing the throwing process by centring the top of his cone of clay before opening out his vessel. This technique of using a pre-made cone of clay has been and is a common technique still in use by potters using the "Japanese style," where the pot is formed entirely at the top of the piece of clay (Cardew 2002, p. 125). This process makes centring easier and enables the potter to utilise the weight of the lump of clay to increase the momentum of the wheel (Rice 1987, pp. 128-129). The system is still in use by the potters of el-Fustat in Cairo where several standard sized pots are thrown from the same lump of clay (van der Kooij & Wendrich, 2002, p. 150).

Potter 4 is at the next stage of the centring process, and has opened out the bowl he is forming by pinching the edge of the clay with his thumb on the inside while keeping his hands on the outside. Note that the potter's wheel must be capable of rotating for some time before needing to be re-spun so he able to use his other hand in shaping the underside of the pot. Potter 5 is similarly engaged with shaping the underside of his pot and is utilising both hands in order to do so, suggesting that sufficient momentum for centrifugal force has been achieved. Potter 6 is trimming and cutting off his finished pot, behind him is a grey restricted vessel.

The final potter 7 (see Figure 3.6), like potter 1, is engaged in shaping the rim of a restricted vessel and, like potter 1, does not have a lump of clay to provide additional momentum to the speed of his wheel as he is most likely engaged in the final stages of finishing his vessel. Generally, when modern potters are finishing and shaping a pot on a modern electric wheel, they significantly reduce the speed for the finishing process (Birks, 1979, pp. 13-5; Cardew, 2002, p. 125; Ruscoe, 1963, p. 185). These two potters may also be representing a secondary finishing phase since after the pot has been left to dry for a time it is placed back on the wheel and trimmed of excess clay or the rim reshaped, which is a process known as *turning* (see Chapter 2).

The vessels, once they have been formed, must be placed to dry and in the second (lower) of the two registers the scene begins with rows of finished pots being placed to dry by an assistant and helpfully it is captioned "drying" (Holthoer 1977, p. 12). The work carried out by the figure behind the assistant, but facing to the right, is not clear because of damage. He appears to be doing something with a lump of clay, though it is possible that it is in fact an unfired vessel in which case he may be trimming it or burnishing it. Adding handles would be another possible action for this figure, however, although the stages in potting are not shown in strict order they do have logic to them and one would expect to see the handle maker before the drying scene.

There is no scene showing the loading of a kiln in the tomb of Bakt III, perhaps because it was deemed unnecessary given that the more colourful unloading scene was

FIGURE 3.7: THE POTTERY WORKSHOP SCENE FROM THE TOMB OF AMENEMHAT. AFTER: NEWBERRY 1893, PG 30-31, PL XI

depicted. It is more colourful because here it is clear that the grey pots have become red. In the unloading scene, a worker reaches into the kiln and passes a red pot to his colleague who stacks similar pots in the space between them. The left foot of the first man is raised showing that the kiln is sufficiently tall to warrant his drawing himself upwards to reach in and take the vessels. To the right of the kiln, a figure with two baskets suspended from a yoke carries away the red, fired, vessels.

The west wall of the main chamber of the tomb of Amenemhet is concerned with daily life scenes and crafts (see Newberry (1893, pp. 30-31, pl XI)). Once again, instead of a logical left-to-right progression, the scene starts with a kiln, goes on to production on a wheel, then clay trampling, another kiln, taking products from the kiln and it finishes with more wheel work (see Figure 3.7).

The clay preparation scene is almost identical to that seen in Bakt III. The two men stand opposite one another each with one leg on the clay and the other on the ground, though in this scene the arm of one passes behind that of the other whereas in Bakt III only the legs are shown in this way. In both instances, this is a rare illustration of perspective in Egyptian art.

The firing and unloading scenes are also very similar to those in the tomb of Bakt III. At the far left of the scene we again see a potter seated in front of the stokehole of a kiln with his left hand raised in front of his face, but this time his other hand is visible and holds a stick. The stick is probably a poker used to spread the fuel within the kiln. It is less likely that it is part of the fuel supply since the artist would have otherwise depicted a heap of fuel to make it clear that this was what was intended. It is interesting that this kiln, and the other shown in the centre of the scene are greyish white in colour. This might suggest that they have been given a coat of lime/gypsum plaster or have been plastered with a highly calcareous clay which has fired to white. This feature has not been observed by the authors on any contemporary kilns (Nicholson & Doherty 2014).

Again in the tomb of Amenemhat, there is no scene showing the loading of a kiln but grey pots are visible in front of the kiln which is being fired. They are probably intended to show that they are being dried around the kiln ready for loading, a phenomenon common to this day. As in the tomb of Bakt III, in the unloading scene a worker reaches into the kiln and passes a red pot to his colleague who stacks similar pots in the space between them. It should be noted that this part of the scene as copied by Newberry is inaccurate. The publication shows the stacked pots in outline (= grey) whilst the pots in front of the worker reaching in are black (= red) suggesting loading rather than unloading. In fact all of these vessels are red. As in Bakt III, a figure with two baskets suspended from a yoke carries away the red, fired vessels. The publication is inaccurate here too since it shows a mixture of red and grey vessels, when all the vessels should be red.

The figures at the wheels also display clear similarities to those of Bakt III. The potters all wear kilts, have cropped hair and are seated in a squatting position, either with one or both knees drawn up towards them. All are engaged in the various shaping, trimming and opening out processes noted in Bakt III. Many of the potters have a large lump of clay on their wheels, and most of them have an almost complete vessel on its top. Potter 1 is engaged with trimming the sides of his bowl with his thumbnail or trimming tool (examples of such tools made of fired pots, bones and shell have been uncovered at the Iron Age Lachish Cave pottery site (Magrill & Middleton, 1997, pp. 68-9,72, fig 6a; Tuffnell, 1958, pp. 291-3, pl 49:12-13 see Chapter 6), with an assistant ready to hand him the next lump of clay. At Fustat, as observed by van der Kooij and Wendrich (2002, p. 147), the master potters relied on their younger assistants to keep them supplied with pre-made wedged humps of clay. Potters 2 and 3 face away from one another and are engaged with removing their vessels from the wheel using a specially designed tool, perhaps comprising a piece of string attached to wood as recorded by Blackman (1927, pp. 152, fig 80). Potter 4 is almost identical to potter 7 in the tomb of Bakt III and is finishing the rim of the pot with the tip of his fingers. A variety of open and restricted vessels are above potters 2, 3 and 4 including cups, plates, spouted vessels, handled jars and pot stands, presumably all fashioned on the wheel and being left to dry on shelves.

Turning to the somewhat later tomb of Khnumhotep (BH tomb no. 3, see Figure 3.8), the craft scenes are also on the West wall of the main chamber though divided by a doorway. Those to the left of the doorway include the potting scenes that occur in the fourth register from the bottom (Newberry, 1893, pp. 68, pl XXIX).

It is uncertain whether the woodcutters are cutting down wood for the pottery kiln or for the boat builders in the next scene. Holthoer (1977, p. 15) has suggested it is more likely for the kiln, and has translated the text as "*srwd*"

FIGURE 3.8: THE POTTERY WORKSHOP OF KHNUMHOTEP III AT BENI HASAN.
AFTER: NEWBERRY, 1893, PP. 68, PL XXIX

'planting' but another reading could be" *swꜣ ḫt*" cutting down trees. The next scene with Khnumhotep being carried in the litter seems quite separate. Newberry places the wood cutting together with the boat building under letters I-K in his publication, (1893, p. 68). Underneath the tree are several gazelle resting, with one reaching up as though to nibble at the leaves as the woodcutters are breaking off the branches. The tree may perhaps be either an acacia or a tamarisk and is therefore likely to be associated with the boat building or carpentry rather than fuel for the kiln and should therefore be viewed as a separate scene (Brewer, Redford, & Redford, 1994). Ethnographic studies suggest that potters use any local garbage as fuel as well as rags, straw etc (Nicholson, 2002, p. 143; van der Kooij & Wendrich, 2002, pp. 150-1). In contrast, at the New Kingdom site of Amarna the most common fuel in the kilns sampled at area O45.1 was acacia (Gerisch, 2007, pp. 169-171, fig A3.1). It is suggested that the Amarna workshops have royal significance so perhaps these potters would have had greater access to higher quality fuels. Do. Arnold (1993, p. 48) postulates an alternative view, by suggesting that the scene is taking place in agricultural land, and that the potter is working for the herdsmen. She postulates that the gazelles are in fact goats, as the pots being produced seems to be milk pots and a dipper juglet.

The potting scene itself is very truncated. A potter is shown at his wheel[20] quite squashed in next to the kiln and a wood cutting scene in comparison to the earlier potting examples. The potter is similarly dressed in a white kilt, but this time covering his upper body in a similar manner to that of the tomb of Ty (see Figure 3.2), with cropped hair as in the other tombs, but this time with a beard. A beard and a larger tunic may represent that this potter was of relatively higher status than the other craftsmen. There are hints of this from the written sources (see next sections) e.g. the potter Sobekhotep was sufficiently wealthy to set up his own dedicatory stela (Ward, 1982, pp. 69, no. 570c).[21] The potter is seated in a crouched position with his knees drawn up towards his body, although he is leaning over in an awkward manner to shape the pot, perhaps indicating that the vessels above him were painted first, or that his arms were made too long. The finished vessels include a set of miniature offering pots on an offering stand, open bowls, jars and small pots. Above the potter, the caption reads *ḳd sꜥnḫ* "the potter or artist fashions." Only the wheel, the covering of the kiln and one offering stand is coloured in black by the copyists (Newberry 1893, pl XXIX), whereas Holthoer (1977, pp. 15, fig 18) retains all of his figures in outline (see Figure 3. and Figure 3.). The wheel appears to be rather small in comparison to the potter and to the restricted rim pot that he is shaping with his right hand while rotating the wheel with his left. The parts of the wheel (wheelhead, pivot, and socket) are distinguished by changes of angle, but not demarcated in colour in Newberry's drawings. The vessel seems to be almost completed. Rather than forming the vessel on the top of a hump of clay, the base of the vessel seems to have been already trimmed, as it tapers from the upper third of the rim of the pot (marked by a line) down to the base. It is interesting, that this scene is the only one in a tomb painting depicting carpenters alongside potters; in contrast, they are commonly put together in wooden models (see next section).

To the left of the wheel an assistant is seen unloading the kiln and the hieroglyphic caption reads, *šdt* "taking away" (Holthoer 1977, p. 15). Only the potter's wheel, the lid of the kiln and one of the pottery stands above the potter is in black (=brown in the tomb), as is the writing. It is unlikely that the pots in the kiln would be grey as they are being *šdt* "taken away" as the caption reads. Unhelpfully, in the other plates, all of the vessels are in outline (apart from pottery stands). Newberry seems to be adding artificial contrast in this volume between the figures and the objects they are carrying. Confusingly, when *Beni Hasan Volume 2* was published in 1894, the pots coming out of the kiln (plate VII tomb 15) are brown, but others are in outline. The kiln itself seems to be having its top opened in some way, perhaps by removing a covering of sherds or, perhaps less likely, clay. As in the Amenemhat scene the worker is using a step to reach into the kiln, thus illustrating its size. Holthoer (1977, p.15) believes the kiln

---

[20] Interestingly he is partly obscured by the wood-cutters, perhaps in an attempt at perspective.
[21] Berlin Museum 12546.

to have "strengthening hoops" around it since it is shown with somewhat wavy horizontal lines. This is certainly a possibility, though Holthoer's figure 18 shows more of these lines than does the copy by Newberry (Newberry, 1893, pp. 68, pl XXIX). Against this view is the fact that the lines are shown either side of the stoke hole which would mean that they could not operate as reinforcing hoops. It may be that this is an attempt to show rather irregular courses on the kiln, though if that is so it is not clear why no vertical courses are shown.

The potters' workshop scene depicted in the tomb of the Nomarch Djehutihotep dates to the reigns of Sesostris II and III (12th Dynasty). It can be found in the east wall of the inner chamber, in association with scenes of harvesting, wine making, preparation of reed mats and bread dough kneading (Newberry & Griffith 1895; see Figure 3.9 and Figure 3.10). These representations are painted in relief, but are unfortunately in a rather poor state of preservation. The scene from left to right shows a possible potter bending over to the left to knead some clay, with text above *kd* "fashioning, or creating," followed by a figure leaning over to the right above a large lump of clay. After this figure are two people seated on reed chairs in front of one-legged tables, holding on to the lip of the wheelhead with the left hand, while using the right to shape a jar. Holthoer (1977, p. 14) has suggested that this scene represents the reshaping of the vessel base into a rounded one by upending it on to the wheel or alternatively the potter could be making the finishing touches to the rim (Arnold, 1993, p. 58).

At the far right of the scene a separate potter is engaged in making bread moulds ("bodega vessels") using a *patrix* or former. This scene is difficult to interpret, as the roughly contemporary scenes at Beni Hasan still show their potters working at the lower pivoted wheels, and seated on the floor or on a block. All of the potters in the scene in Djehutihotep's tomb are seated, so one wonders whether this is meant to represent their status as a craft worker rather than the way that they were actually undertaking their work. The design of the wheel that they are using is very similar to the Beni Hasan design, with a socket and a pivot but elongated axis in Djehutihotep's tomb to accommodate the seated position of the potter. The more complete representation of the seated potters shows that the wheelhead is attached to a plug-shaped item, which could be the pivot attached to the wheelhead and then slotted into the socket (see Figure 3.9). The wheel has a very high axis which would not be able to support the combined weight of a clay wheelhead (56-93 Kg, (Powell, 1995, pp. 320-1)) and stone pivot wheel bearing (some pivots weigh up to 7kg see Chapter 2, Table 2.1) without the use of a bench, something that is not depicted in the scene. This scene, rather like the one found in the tomb of Khentika (see Figure 3.4) seems to represent the potter's wheel as a spinning top, which would be awkward to keep upright and to prevent from falling off its socket, possibly falling over when the wheel slowed down (Powell, 1995, p. 318). Alternatively, this scene could simply represent something similar to the offering stand shape, being used as a finishing stand which the potters are using to finish their pots without rotating them, rather like those in later

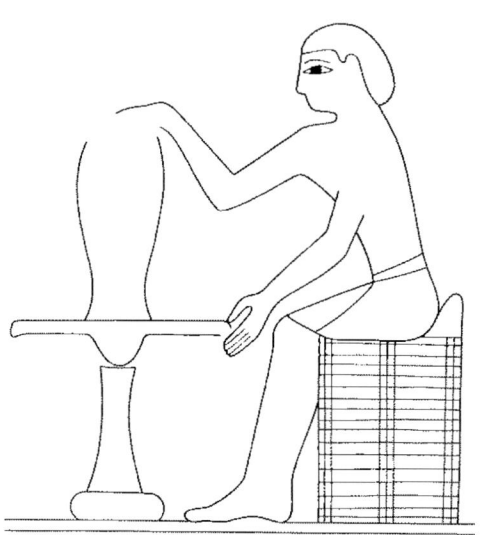

FIGURE 3.9: THE MOST COMPLETE POTTER REPRESENTATION FROM THE POTTERY SCENE FROM TOMB OF NOMARCH DJEUTIHOTEP, DEIR EL BERSHA (AFTER: NEWBERRY & GRIFFITH 1895, PL. 25)

FIGURE 3.10: THE POTTERY WORKSHOP SCENE IN THE TOMB OF DJEUTIHOTEP. NEWBERRY & GRIFFITH 1985, PL 25

FIGURE 3.11: THE POTTERS FROM THE TOMB OF HOREMKAWEF, HIERAKONPOLIS, SECOND INTERMEDIATE PERIOD. AFTER FRIEDMAN 2006, PG 25

temples and *mammisi* where Khnum is generally shown making the final touches to the ready-formed clay Pharaoh e.g. Temple of Hibis (Davies, 1953, pl 27). It seems most likely that these potter's wheels are represented without a supporting bench or other structure, as found in ethnographic studies of potters, otherwise it would be too unstable (see http://bit.do/potterswheel of wheel throwing scenes at El Nazla pottery).

The tomb of Horemkawef re-recorded at Hierakonpolis by Friedman (2006, p. 20) contains the image on the western wall of a potter working at a wheel with help from an assistant (see Figure 3.11). Both men kneel either side of the potter's wheel. The man on the left is labelled "potter"[22] and is shaping a round-bottomed slender-necked jar with his right hand, while inserting his left into the mouth of the pot, through which the potter's hand can be seen. The potter's assistant is engaged in steadying or spinning the wheel, the first depiction of an assistant. It is difficult to make out the particulars of this scene from the publication, but it appears that the vessel is already formed, and has been placed on the top of a rather large lump of clay while the potter shapes the mouth. This might represent two-stage or three-stage throwing.

The final potter's workshop scene to be recorded in a tomb occurs during the reign of Amenhotep II (1427-1401 B.C.) in the mid 18th dynasty. The tomb of Kenamun at Thebes (TT 93) also shows a similar scene to the one in Horemkawef, with a potter and an assistant at work together at one wheel (see Figure 3.12). The potter uses his foot to steady the wheel, while his assistant grips the wheelhead. The potter's wheel is depicted with only the pivot showing, and no socket. The pivot seems to be balancing precariously on the ground as the potter has

a tower of clay, half the size of himself, in a large cone wobbling on the wheel. If the drawing by Davies (1930, pl 59) is to be believed, the clay is not attached securely to the wheel-head, and is leaning towards the seated potter with quite a large gap of air between clay and potter on one side. It is likely that such a great amount of clay placed on the wheel would indicate that the potters were creating large vessels similar to the round bottomed or tall-necked storage jars depicted in the scene.

So far, only the artistic representations of potters have been taken into account as they were represented in tomb scenes. None of these tombs were designated for the potters themselves. Rather, the potters are depicted alongside a variety of other craftsmen and women, or as part of baking and brewing scenes. The potters were not, of course, responsible for the constructing the scenes in which they were depicted, this would have been under the jurisdiction of the funerary artists in consultation with the tomb owner. However, the stone carvers and painters who were designing the tomb scenes obviously had a clear understanding of the pottery making process, as they appear to have recorded pottery manufacture in some cases extremely accurately (Nicholson and Doherty 2014). In some cases, it is the modern copyist at fault rather than the ancient artist. The main discrepancy seems to be their depiction of the potter's wheel, which is quite different in each of the cases mentioned above. The possible reasons for this will be discussed in Chapter 7.

In the next sections, other secondary sources of depicting potters will be investigated. For a short time during the Old and Middle Kingdoms, it became common for Egyptians to represent some tomb scenes in the form of three-dimensional models rather than on tomb walls. Offering bearers, soldiers, granaries, craft workers, boats and many other activities came to be wrought in wood, stone and occasionally in ceramics (Breasted, 1948). Infrequently, potters were also depicted in this manner, and some examples are next presented.

Potters at their Wheels: Wooden Models and Limestone Statuettes

Further secondary evidence for the use of the pivot and socketed wheel is derived from a small limestone statuette, reputed to be sourced from the 5th dynasty tomb of Nikauenpu in Giza (see Figure 3.13; Breasted, 1948, pp. 49, pl 45). This statuette seems to depict a potter shaping a vessel in a similar manner to the one in the tomb of Ty (see Figure 3.2). The wheel socket appears to be embedded into the ground, with the pivot placed on top and a wheelhead attached to the pivot by means of clay. The limestone is painted a mud brown colour, suggesting the ubiquitous application of clay by the potter to his wheel, apart from his white kilt, and the slightly darker brown ground which he is separated from by a low stool. Note that as in the tombs scenes, the statuette of Nikauinpu is seated on a block rather than on a stool or chair, suggesting that in the 5-6th dynasties at least, potters sat and worked at their

---

[22] Friedman does not detail the hieroglyphs, but one would assume that the caption reads *ḳd* as building or forming.

FIGURE 3.12: POTTERY WORKSHOP OF KENAMUN (TT 93), THEBES. AFTER: DAVIES 1930, PL 59

FIGURE 3.13: SERVANT STATUETTE OF POTTER, PERHAPS FROM 5TH DYNASTY TOMB OF NIKAUINPU, GIZA [E10628] 13.2 x 6.7 x 12.5CM. PHOTO: ORIENTAL MUSEUM COLLECTIONS, CHICAGO

There are a variety of wooden models dating to the First Intermediate Period and Middle Kingdoms known to contain scenes of potters working at their wheels, see Table 3.1 (two examples from the tomb of Karenen (Quibell, 1908, pp. 10-11, 75-6, pl 17 1,3 & 19,4), one in the tomb of Gemniemhat (Firth & Gunn, 1926, pp. 53, pl 29 C), the tomb of Inpuemhet and Usermut (Quibell & Hayter, 1927, pp. 40-41, pl 24), and from the tomb of Pharaoh Montuhotep II (2061-2010 B.C.) (Arnold Di, 1981, pp. 33, pl 37). These have been variously described by Breasted (1948, pp. 49-51) Holthoer (1977, pp. 10-11, 15-16) and Do. Arnold (1993, p. 69) has postulated that these wooden models represent another type of pottery wheel the "extra low simple wheel." However, when one examines these models, it appears that what is being represented is in fact a 3D version of the scenes depicted on tomb walls. The potters all sit on the ground or on a block with their knees drawn up to their body. With their right hand they shape or throw the vessel and with their left they spin the wheel, with a water pot nearby to moisten the clay. Often they are sitting near to a kiln with an assistant close by making up fresh cones of clay to be later applied to the wheel so that the potter can continuously throw pots in the manner of an assembly line (see Figure 3.6, potter 2 and Figure

wheels in this manner, and it was not until later (post 12th dynasty that stools or chairs were used) and even then only for particular jobs e.g. working large amounts of clay as shown in the tombs of Kenamun (Davies, 1930, pl 59) and Horemkawef (Friedman, 2006, p. 20).

From the late Middle Kingdom, a statue reputed to be a potter called *sebkhotp* provides some information regarding the title "potter".[23] His statue contains the standard offerings for bread, beer and fowl, but uses the words *ikdw ndst* suggesting that this was meant to signify a potter rather than a builder as this phrase literally means "builder in little."

FIGURE 3.14: A CLOSE UP OF THE POTTER'S WHEEL IN GEMNIEMHAT'S TOMB AT SAQQARA, AEIN 1633 ©NY CARLSBERG GLYPTOTEK. PHOTO: IVOR PRIDDEN

---
[23] Berlin Museum 12546.

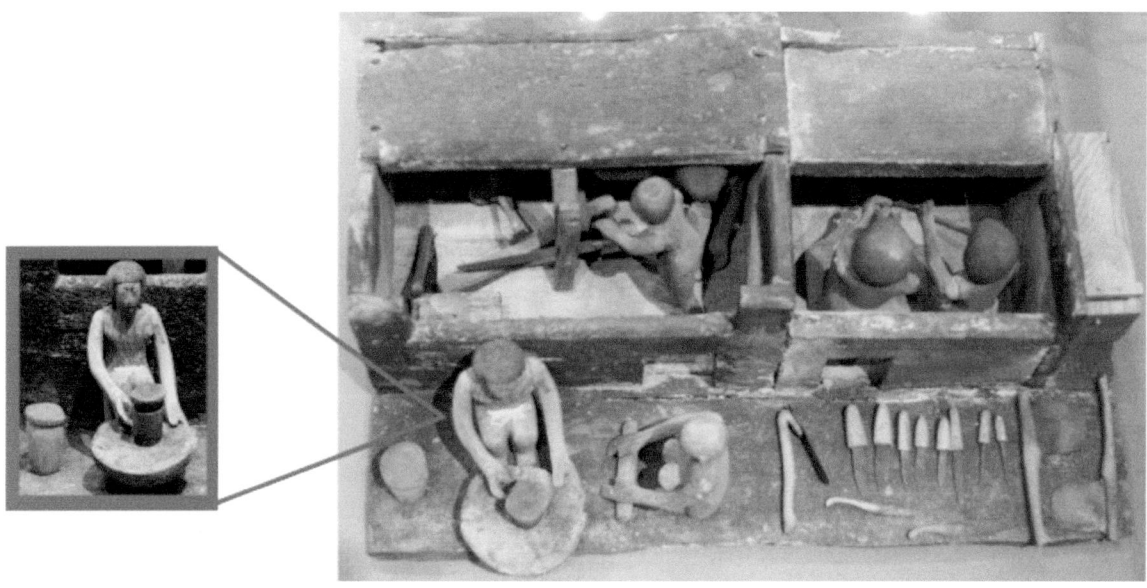

FIGURE 3.15: WOODEN MODEL FROM THE TOMB OF GEMNIEMHAT AT SAQQARA, AEIN 1633 ©NY CARLSBERG GLYPTOTEK. RIGHT: BIRD'S EYE VIEW SHOWING CARPENTERS AT WORK BEHIND THE POTTERS IN THE PARTIALLY COVERED BUILDINGS, WHILE THE POTTERS WORK OUTSIDE. ONE WORKS THE WHEEL; THE OTHER PREPARES FRESH CONES OF CLAY TO PASS TO THE OTHER POTTER WHEN NEEDED. THE TOOLS ARE FOR THE CARPENTERS. INSET: FACE VIEW OF POTTER WORKING A POT ON HIS WHEEL, LEFT HAND SPINS THE WHEEL, RIGHT WORKS AT SHAPING THE POT. NOTE THE COMPLETED POT OR POSSIBLY A WATER JAR TO HIS RIGHT. PHOTOS: IVOR PRIDDEN

Model location and details:	Date:	In association with:
Karenen model 1, Saqqara, Egyptian museum 39131	First Intermediate Period (reign of Amenemhet I)	Carpenters (in a separate roofed workshop)
Karenen model 2, Saqqara Egyptian museum JE 39132	First Intermediate Period (reign of Amenemhet I)	Potters alone, outside
Gemniemhet Saqqara, AEIN 1633 Ny Carlsberg Glyptotek, Copenhagen	First Intermediate Period (reign of Amenemhet I)	Carpenters and blacksmiths (in two separate roofed workshops), potters outside
Model of Inpuemhet and Usermut Saqqara, Egyptian Museum JE45319	First Intermediate Period	In one workshop with carpenters, stone vase maker, fire heater (possibly a metal worker). Roofed on one side to cover box containing carpentry tools).
Models (at least two) from tomb of King Mentuhotep II, Deir el Bahri. British Museum BM47655	First Intermediate Period, c.2061-2010 B.C.	Potters alone

TABLE 3.1: WOODEN MODELS OF POTTER'S WORKSHOPS AND THEIR DETAILS

3.14). The wooden model wheels are very similar to those depicted in the statute of Nikauinpu's potter (see Figure 3.13 and Teeter 2003, pp. 25).

Wheel-heads are shown as thick relative to the size of the model men (e.g. Montuhotep's model wheel-head BM47655 was 6.7cm in diameter and 0.9cm thick and see Figure 3.15). They are usually not completely circular and attached to the model floor with a peg, possibly made to resemble the pivot and socket bearings. Often the wheelheads have traces of red paint; possibly to look like fired clay and the pots have traces of black to signify damp clay. The model from the tomb Gemniemhat, Saqqara,[24] (Arnold, 1993, pp. 69-73, fig 84, 86 A-B; Breasted,

1948, pp. 51, pl 46b; Firth & Gunn, 1926, pp. 53, pl 29C; Holthoer, 1977, pp. 11, fig 13), provides a perpective of how the workshops would have been organised (see Figure 3.15).

The number of wooden models depicting potters so far discovered are relatively few (approximately six, see Table 3.1 above). It is interesting to note that in many cases the potter's workshop is beside a carpenter's and with at least one stone vase driller, perhaps signifying that the crafts were linked in ancient times. Shaw (2004, p. 16) has suggested that industrial workshops may not always have been buildings at all and that many craft activities would have taken place in open courtyards; these models may provide evidence for this proposition in relation to potters. It would make sense for at least some of the

[24] In the Carlsberg Glyptotek AEIN 1633.

potters' activities to occur out of doors, and as the models indicate, perhaps wheel throwing and kiln firing were such actions. Many of the model workshops are partially roofed, presumably suggesting that roofs were needed to keep off the heat of the day, but with the majority of the industrial processes taking place in the open air. It is perhaps significant that the later 12th dynasty tomb of Djeutihotep contains an apparently different type of potter's wheel with a "tall stem" (see Figure 3.10) whilst the even later Second Intermediate Period tomb Horemkhawef depicts the same as the First Intermediate Period wooden models (see Figure 3.11), Beni Hasan tombs (see Figures 3.6, 3.7 and 3.8) and 18th dynasty tomb of Kenamun (see Figure 3.12). This suggests that Djeutihotep is the anomaly, rather than an indicator of changes in potter's wheel technology.

So far, the representations of potters depicted on tomb scenes, in stone statuettes, and in wooden models have been described and interpreted. In the next section, the written sources of the potter's wheel will be explored, including literature, *Pyramid Texts*, lists and other areas of interest. By examination of all the evidence of the potter's wheel one can gain understanding of its origins and use in Egyptian society.

Written Evidence for the Potter's Wheel

To date, there is only limited written evidence of the first use of the potter's wheel. The basic word for potter comes from the verb "*qd*" or "*kd*" which has a variety of meanings: 'to build', 'to create', 'to form' or 'to fashion'. The sign is not exclusive to potters, but was often used as a general term for builders or craftworkers, and relied upon the use of a determinative or tomb illustration to signify that the text was implying pottery making rather than anything else. In fact, pottery workshops depicted in tombs are rarely accompanied by captions, as the verb in the scenes is determined by the actual potter depicted. Only four such scenes are captioned "*kd*" which are known to the author; (1) those of Ty at Saqqara (Épron & Daumas, 1939, pl 71; see Figure 3.2), (2) Ptahshepses at Abusir both dating to the 5th dynasty (Vachala, 2004a; 2004b), (3) Djehutyhotep at Deir el Bersha dating to the 12th dynasty (Do. Arnold, 1993, pp. 59, fig 67; Newberry & Griffith, 1895) and (4) Khnumhotep III at Beni Hasan also dating to the 12th dynasty (Holthoer, 1977, p 15, fig 18; Newberry, 1894, pl XXIX). The word "*kd*" is much more likely to be associated with building or creating than pottery making.

The sign is often attached with the plasterer's float and the phoeneme "d" and a variety of determinatives, usually a circle, such as in the tomb of Ty (see Figure 3.2). From the Middle Kingdom, the circle determinative is replaced by the *nw* pot, occasionally accompanied either a quail chick or coil of rope for the *w*. The word "*kd*" is often with a brick enclosure wall and sometimes with a mason working on it and is therefore associated with builders, brickmakers and building. The word is often connected to industry, craft occupations, the manufacture of statues or gods, divine birth and creation (Dorman, 2002, p. 83). From the 18th Dynasty of the New Kingdom and later the potter's wheel is used as a determinative in the "divine birth" rooms (Davies, 2004) and *mammisi* (Kockelmann, 2011, p. 5) in temples, and related to concepts of the birth of both mortals and gods in human form. Odler (in press) has highlighted possibly the earliest uses of potter as a hieratic determinative from the papyri found amongst the finds from an anonymous tomb in Gebelein, Upper Egypt (Posener-Kriéger 2004, p. 13). Three signs dating to the 4th dynasty seem to depict potters, each slightly different (possibly different scribal hands at work); once on verso of the third papyrus and two times on the fifth papyrus (Rocatti, 2006, p. 87). However, these documents do not describe potting or potters, but instead use the figure of a potter as a determinative for the word "*kd*" in relation to building or creating.

The title "potter" together with the potter's name appears in a caption from a scene of the 5th dynasty (probably during the reign of Niussere (Krejčí, 2000; 2009, p. 145)) found in the tomb of the Vizier Ptahshepses[25] (Vachala 2004b, pp. 176-9; see Figure 3.16). It reads: *p3 ikdw n pr dt Wri* translated by Senussi (2006, p. 329) as *"The Potter of the House of Eternity (cemetery) Weri,"* with *ikdw* signifying the term potter. Alternatively, *pr dt* could also be translated as mortuary estate (Warden, 2010, pp. 185, note 4). Senussi (2006, pp. 329-30) has cited this depiction as the first representation of a kick wheel. Odler (in press) has suggested that this is a representation of a seated potter throwing on the wheel in a similar manner to the hieratic archival document of Raneferef (see Figure 3.17) or a representation of a potter at a "tall stemmed" wheel in the manner of the scene in Djheutyhotep's tomb. In the author's view, neither of these suppositions can be right. The rest of the scene does not depict a potter working at a potter's wheel, rather it illustrates the manufacture of beer jars, the firing of the beer jars and in the registers below, the filling and sealing of beer jars. To the author, the caption "potter" (see Figure 3.16) rather than representing a kick wheel or hand spun potter's wheel is, in fact a seated figure making a beer jar by hand, with the beer jar resting in a chuck or some other support. Senussi (2006, p. 330) nonetheless considers that the potter in the caption is making a beer jar, but on the kick wheel. Although most of the evidence indicates that beer jars were not made on the wheel during the 5-6th dynasties (see Figure 3.18 and below) and that the kick wheel is a much later introduction. Since the kick wheel has not been sourced archaeologically at this date in Egypt and not until possibly the Late Period, it is unlikely to be represented.[26] The scene is highly fragmentary and is therefore open to interpretation, but as the rest of the

---

[25] Fragmente 57(B) + 81 + 93 + 221 (Vachala, 2004a).

[26] So far the advent of the kick wheel has been dated to c.500 B.C. based on a representation of a foreign looking Khnum working a kick wheel at the Persian temple of Hibis in the Dakhla Oasis (Arnold, Do. 1993, pp. 79, fig 93A). Petrie also found possible potter's bats dating to the 7-6th C B.C. at Tell Dafana, Eastern Delta (Petrie, 1888, pl 34, nos 35 & 36 inverted), but the earliest use of the kick wheel has yet to be determined.

FIGURE 3.16: RELIEF FROM THE TOMB OF THE 5TH DYNASTY VIZIER PTAHSHEPSES *PA ı͗ḳdw n pr dt WRI* "THE POTTER OF THE MORTUARY ESTATE, WERI" *ı͗ḳdw* "POTTER" IS CIRCLED. AFTER VACHALA 2004A, P. 179, FRAGMENTE 57(B)+81+93+221

FIGURE 3.18: AN EXAMPLE OF A HANDMADE 6TH DYNASTY BEER JAR BUILT BY COILING (SEE TOMB OF TY FIGURE 3.2) FROM SAQQARA SQ98477 K 98-195. AFTER RZEUSKA 2006A, PG 60, PL 19, PHOTO 69.

scene depicts the making of handmade jars, it is more than likely that this is also the intention of the caption. It is quite striking that this scene is very similar to representations of the word "potter" in the *Pyramid Texts* that also seem to be the hands and heads of potters making beer jars (see Figure 3.19).

The title "potter" also appears on a 6th dynasty hieratic clay tablet from Balat, in the Dakhleh Oasis, where it is used to determine the word potter "*ikdw*". The text records that the potter had not yet come to the place called *Rwd.t* where he was charged to prepare the journey for the chieftain of *Dmj-jw* (Odler, in press; Pantalacci, 1998, pp. 303, fig 1). There is some fragmentary evidence surviving from the administrative archives of the mortuary temple complexes of Raneferef Isi, his father Neferirkare Kakai and Queen Khentkaus II at Abu Sir (Posener-Kriéger & de Cenival, 1968). The Neferirkare archive dates from the reigns of Djedkare Isesi of the 5th Dynasty to Pepi II of the 6th Dynasty (c.2300-2181 B.C.). These archives provide a brief glimpse into the economic life of the funerary cults as sources of revenue and the large amounts of food (bread, beer, oxen and birds) delivered from the Ptah Temple in Memphis to Abu Sir (Posener-Kriéger, Verner, & Vymazalová, 2006). A potter's workshop is suggested to be located near the cult temple of Neferirkare indicated with the words '*rr.t n.t nhp*" literally the '*rr.t* of the potter's

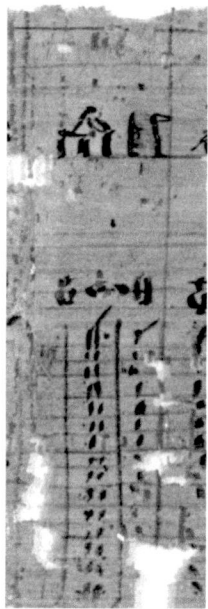

FIGURE 3.17: SECTION OF PAPYRUS FROM THE ARCHIVE OF THE RANEFEREF'S MORTUARY TEMPLE SHOWING THE INSCRIPTION *Ḳd NṬR* AT THE TOP OF THE COLUMN. AFTER POSENER-KRIÉGER, VERNER &VYMAZALOVÁ 2006, PL 49.

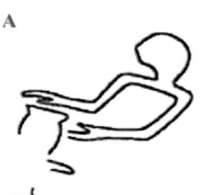   

FIGURE 3.19: THE PYRAMID TEXT REPRESENTATIONS OF POTTERS (L-R) A-C FROM PYRAMID OF PEPI I (LECLANT, MATHIEU, & PIERRE-CROISIAU, 2001, PL XVI: COL 8, XXIV: COL 30), D FROM THE PYRAMID OF MERENRE (SETHE, 1910, P. 160) FROM THE PYRAMID TEXT SPELL 1185

wheel, postulated as the written evidence for the possible pottery workshop located within the temple of queen Khentkaus II. It is the earliest written evidence of the word *nḥp* "potter's wheel."

A second document from the papyri archive of Pharaoh Raneferef at Abusir, contains the title of "potter of god" *ḳd nṯr* or the "divine potter" and is included in the list of the workers of Raneferef's mortuary temple. In return for his work, the potter received one loaf of bread *ḥt3* and two loaves of *psn* and rather sadly, no jars of beer *ḥnḳt* unlike some of the other workers in the same list e.g. gardener (see Figure 3.17 and Figure 3.18).

This scene has also been viewed as a representation of a stand or potter's wheel, however, it could also be a hieratic misnomer, where the scribe by seating the potter was trying to emphasise the potter's status (Odler in press). In the case of the Raneferef document, the "divine" potter appears alongside a bleacher, gardener, keeper of cloth, cook, physician of the Great House and a craftsman, who all receive similar amounts of bread and beer (Posener-Kriéger, Verner, & Vymazalová, 2006, pp. 266-268, pl. 48-9). These authors have suggested that rather than being viewed as a potter's wheel, the hieratic should be viewed as a potter's stand A158 (Posener-Kriéger, Verner, & Vymazalová, 2006, p. 442).

The *Pyramid Texts* also make use of the determinative "potter," in the so-called ferryman texts, but depending on the translation, the word can signify creating or potting. Utterance 516 line 1185-6 from the *Pyramid Texts* of Pepy I, Merenre and Pepy II is translated by Faulkner (1969, p. 190) as "I am your potter/creator upon earth who broke the egg (?) when Nut was born. I have come and brought you this mansion of yours which I built for you on that night when you were born, on the day of your birth place (?); it is the beer jar (*ḏwiw*)".

In the determinative examples A-C (see Figure 3.19), the potter involved in spell 1185 (quoted above) is unlikely to be making beer jars on the potter's wheels (see Figure 3.18), as beer jars of the 6th dynasty were only handmade, being coil built and shaped through pinching (Rzeuska, 2006a, pp. 60-102, pl 9-34; Rzeuska & Kuraszkiewicz, 2011, p. 830) so the writers of the text would have probably only been using the determinative in its more figurative sense. In the pyramid texts, it was common practice to display human figures and animals incompletely since it was considered that if the being were whole, they would be able to undermine the Pharaoh's power (Pierre, 1997, pp. 355-360). Thus the potters are only depicted with their heads and arms holding or working a vessel on their potter's wheel (see Figure 3.19). This is unhelpful in terms of trying to understand the seating position of the potters, whether they were kneeling on the ground or seated on a chair, as is the case in the later tomb depictions of Djehutyhotep at Deir el-Bersha dating to the 12th dynasty (Arnold, Do. 1993, pp. 59, fig 67; Newberry & Griffith, 1895). Alternatively,

if the wheel was already well established by the time the texts were written down, the Egyptian artists recording the *Pyramid Texts* might have assumed that most pottery was made on a wheel by that time.

In other places, it is not obvious why a potter was used as a determinative. In many cases in the *Pyramid Texts*, although a potter does appear as the determinative for "*qd*" or "*ḳd*" the actual writing does not necessarily refer to a potter or to the creation of pottery, but rather the word is used in the more metaphorical sense of creation or building. Thus in Utterance 324, line 524 it reads, "Hail to you Khnum, being driven off! May you refashion me." Faulkner (1969, p. 104) referred to his translation of the spell as "a strange sentence" where the "context is quite obscure" and considered that the reading with the potter determinative should be *iḳd.k* rather than *iḳd.f* hence his "you refashion me" translation (Faulkner, 1969, pp. 104, note 17-18). However, it could also be translated as "may you create me." Generally, the word "*qd*" or "*ḳd*" refers to building an architectural structure in all periods of Egyptian history, both in private and royal tombs and temples as well as in religious and literary texts. It also can refer to the building of brick or stone.

Texts of the Middle Kingdom begin to use the potter's wheel in a more secular or metaphorical sense, such as in the Papyrus Leiden I344 (*The Admonitions of a Sage*), recto 2:8 in which it states, "*iw ms t3 ḥr msnḥ mi irt nḥp*" "For the land turns round as does a potter's wheel" (Gardiner, 1909; Lichtheim, 1975, p. 151). In other cases, the status of the potter is openly mocked such as in the various oastraca and piecemeal versions of the Papyrus Sallier II, column V, lines 5-6 or the *Satire of the Trades* (Holthoer, 1977, p. 18; Lichtheim, 1975):

Ỉkdw nḏst ḥr 3ḥt "a potter is under (i.e. carries) clay"

ʿḥʿw.f m ʿnḥw ḥmʿ.n sw r. šʿw r š3iw r ps(t) st (3) ḥr 3ḥt.f "His life is like that of an animal. Dirt covers him more than a pig to burn under his earth"

ḥbswt.f nḥt m dbn ʿgs.f m stp "His clothes are stiff from dry clay, his loin cloth is like a rag"

3k t3w r fnd.f prw (m) t3.f wḏ3 "fumes enters his nose directly from the furnace"

Ỉw.f ḥr tity (?) m rdwy.fy shmw im.f ds.f "He tramples with his two legs being crushed by it himself"

ḥmʿ ḥ n pr nb ḥii ny n3 n iwywt "smearing the courtyard of every house and struck is the public places"

The New Kingdom Papyrus Lansing 4:4-5 is similarly derogatory towards potters, "The potter is smeared with mud like a person whose folk have died. His hands and his feet are filled with clay. He is like one who is in the mire." (Gardiner, 1937, p. 103f; Lichtheim, 1975, p. 169). Such texts are meant to put the scribal profession above all others, and denegrade industry and craft, over-emphasising the dirt, heat and smell. However, as the tomb scenes, models and later chapters will indicate, the potting profession certainly was not a clean one, and being near to a kiln was likely to be associated with many unxious smells and black smoke (Nicholson, 1995b).

A stele erected by Ramesses II in Mansiyet es-Sadr praises the work of craftsmen who were engaged by him to cut stone statues throughout Egypt. In lines 16-17, writing to the stonemasons, it mentions potters "I have put many people to provision you against decay, fishermen to bring fish, moreover, vintagers to make ḥsbt, fashioners of ḏs vessels on a potter's wheel (iḳdw n ḏsw ḥr nḥp) making ḥnw-vessels to cool water for you in the summer time" (Valbelle, 1985, p. 148). This perhaps could be indicative of the potter's status, placed together with the providers of foodstuff to the stonemasons (Holthoer, 1977, p. 23), in a similar manner to the lists of the Raneferef archive and in association with baking and brewing scenes (Posener-Kriéger, Verner, & Vymazalová, 2006, pp. 266-268, pl 48-9).

Summary

Now that wide varieties of sources have been consulted regarding the beginnings of the use of the potter's wheel, some understanding has been gained regarding its use. From the authenticated tomb wall scenes dating to the 5th-6th dynasties, it is evident that some potters were attached to estates of Egyptian royalty and nobility e.g. Ty, which also pushes back the date of the first use of the potter's wheel in Egypt than originally suggested from the physical evidence. There is a range of different types of potter's wheels depicted. The scenes are remarkably well detailed, with clear steps in the manufacturing process and firing outlined.

However, the paucity and infrequency of this secondary material requires caution in its analysis; one cannot assume that all these depictions are accurate, or that such activities occurred in a standardised manner throughout Egyptian history; representations of the potter's wheel in texts, tombs, and models are rare and sometimes separated by hundreds of years. There is no pictorial evidence to support the use of the potter's wheel prior to the 5th dynasty, which supports Do. Arnold's (1993) postulation that the potter's wheel was not in use until that date. However, as Chapter 2 has demonstrated, there is a wide variety of archaeological sources for the potter's wheel being in use from at least the 4th dynasty e.g. physical potter's wheel bearings have been located in various sites. Despite this caveat, as has been demonstrated in this Chapter, tomb scenes and texts can provide a wealth of information towards the understanding of the potter's wheel, its introduction into Egypt, and gives hints as to how it might have been used (see Chapter 6). The statuettes and models dating from the 5th dynasty in particular are very similar to those depicted on the tomb walls. The wooden model workshops dating to the First Intermediate Period are very detailed and suggest the use of tools to aid the throwing process. Such scenes introduce the possibility of evidence for apprenticeship and potters working in close proximity to other craft workers, notably carpenters. The written manuscripts dating from the 4th-6th dynasties provide evidence for the first written evidence for the potter's wheel.

However, as yet, the Near Eastern evidence has to be consulted, and this is the choice of topic for the next chapter. Excavations in the Near Eastern have concentrated more on settlements rather than tombs and so have been able to uncover evidence of the industrial and craft working areas to a greater extent than in Egypt. However, the archaeologists who concentrated on the tombs of Egypt have opened up a range of secondary evidence in support of the use of the potter's wheel, as has been illustrated in this and the preceding chapter. The next section will consider the use of the potter's wheel in the Near East and compare it to the evidence illustrated in this chapter regarding Egypt.

Chapter 4:

Inventing the potter's wheel

"Choraebus, the Athenian, was the first who made earthenware vessels; but Anacharsis, the Scythian, or according to others, Hyperbius, the Corinthian, first invented the potter's wheel" Pliny the Elder, *The Natural History*, Bk. 7.57-8

The quote above might suggest that the potter's wheel originated in Corinth or Scythia, and is attributed to a particular individual, however, other myths regard the inventor of the potter's wheel as a specific person in antiquity e.g. the Biblical Adam as the first potter (Genesis 2:7) or Chinese Emperor Huang-Ti (Johnston, 1977, p. 175). Some anthropologists have indicated that they consider that there was a single inventor using the term of "individual innovator." Since stability in technology tends to be the rule rather than the exception that view needs to be explained while taking into account the changes in technology (Foster, 1967). Generally, stability is the normal state for most societies. In reality, however, the invention of the potter's wheel is likely to have been a cumulative process developed over time in the city state workshops of the Near East. In this chapter, the contemporary societies within the Near East and Egypt will be assessed through analysis of their social, economic, political, religious and technological stand points in order to ascertain why they developed the potter's wheel. The precedents and requirements for such an invention will be considered through application of technological theories and investigation of the evidence for craft production and the pottery industry in both situations.

Understanding the Uptake of a new technology: why the potter's wheel?

In this section, an analysis will be undertaken to establish why cultures adopt, adapt, or invent new technologies and techniques, and consideration will be given to the underlying social processes that instigate them. Different societies adopt and adapt a technological innovation in many ways, depending on their own cultural mores and value systems (Patrik 1985, p. 27-62). Therefore, technological innovations have to be understood not just in terms of the artefacts themselves, and how they were invented and made, but in terms of the people whose thought processes enabled them to invent the artefact in the first place, their belief that the artefact could be useful, and how they instilled this belief within their communities. Social and technological innovations are deeply intertwined in the construction of durable cultural stability and structure and should therefore be studied together (Dobres, 2000). A study of how societies have gone about inventing a new technology will firstly be undertaken, and then how the various social, cultural, and political factors aided the instigation of the use of the potter's wheel will be considered, both in Egypt and in the Near East.

Theories of Technology

"the facts of nature form the warp, man's [sic] imagination and inventiveness the woof of the tapestry of our material civilisation," Forbes (1958, p. 6).

When formulating a theory of technology, archaeologists and anthropologists alike have often restricted their thinking to a physical description of how the technology was undertaken. In doing so they have taken into account the constraints of the innovators' environment and postulated that the causes of technological change are due to natural factors rather than human e.g. lack of raw materials (Binford, 1965). These theorists believe that it is the idea, rather than the artefact, that is culturally significant (Foster, 1959b, p. 99). The idea behind the potter's wheel needs understanding, given that the Egyptians were not usually sufficiently driven to take on a new technology when they already had well-established means of manufacture. The Egyptians used relatively few machines in their industries; the exceptions being the lever, the loom (Vogelsang-Eastwood, 2000, pp. 270-1), the pot bellows (Ogden, 2000, p. 157; Nibbi, 1987; Davey, 1979), the twist-reverse-twist stone drill (Stocks, 2003, p. 17), the plough, the waterwheel, the lathe (Gale, Gasson, Hepper, & Killen, 2000, pp. 357, fig 15.21) and survey equipment such as the plum bob and set square. They preferred instead to rely on high numbers of workers performing the tasks. If they wanted to increase production, they simply multiplied up the numbers of workers (Gillings, 1982; Shaw, 2004). Prior to the use of the potter's wheel in the Pre and Early Dynastic Periods (c.4000-3500 B.C.), pottery was being produced on a large scale by specialists for funerary contexts. Most tombs in early cemeteries contained at least one pottery vessel, but some graves contained hundreds e.g. Naqada II cemetery T, demonstrating increasing wealth differentiation and using pottery vessels and grave goods in order to do so (Bard, 1994; Midant-Reynes, 2000, pp. 47-53).

Wendrich (2006, p. 267), separates "techniques" from "technology" by using the former to describe the methods through which raw materials are made into objects and the latter as the knowledge that the craftspeople require to utilise a technique (Richter, 1982, p. 8). These behavioural techniques should not be studied in isolation but should be seen as a means of understanding the social constructs that enable the technology to be undertaken, whether that is the social relations that underlie production, or the laws of

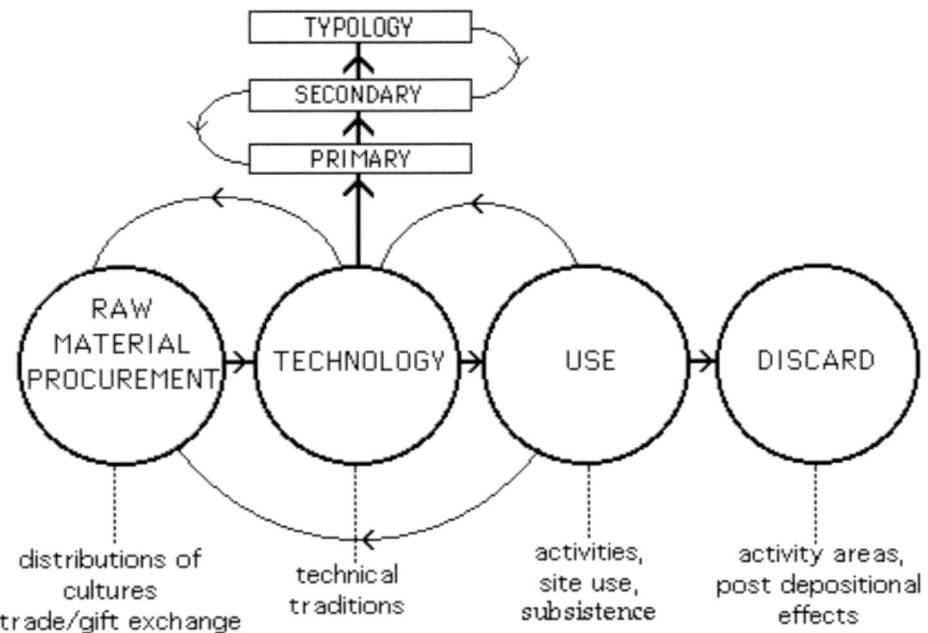

FIGURE 4.1: THE CHAÎNE OPÉRATOIRE APPROACH. THE FOUR BASIC LINKS ARE RAW MATERIAL PROCUREMENT, TECHNOLOGY (SEPARATED INTO PRIMARY AND SECONDARY REDUCTION AND TYPOLOGY), USE AND DISCARD. AFTER: GRACE (1989, P. 3)

matter and energy that form the framework for such social behaviour. Often, changes in technique (such as changes in flint knapping techniques e.g. invention of Acheulean hand axes after the older Olduwan styles) offer insight into social changes. The two have to be studied symbiotically in order to achieve a greater understanding of the society as a whole.

The relationships between technology and society have often been studied independently (van der Leeuw, 2002, pp. 238-240), with either the study of the effect of technological systems on culture and society being the focus, usually as constraining factor (Goody, 1971) or an analysis of how humans communicate when they make or use artefacts e.g. studies of style (Kroeber, 1957). This approach, although useful, ignores that both techniques and technologies are inherently a product of society and should be studied as such, a theory first propounded by Leroi-Gourhan (1943/5). Lemonnier (2002) postulated that the techniques should be studied in their own right, in a "chaîne opératoire" (see Figure 4.1) i.e. a series of operations that transforms one object from a raw material into a manufactured product that is both culturally meaningful and a functional object (Pelegrin, Karlin, & Bodu, 1988, p. 56).

This enables the social control of these various operations to be studied, together with the conceptual and intellectual dimensions of the technology involved and their role in society (Lemonnier, 1980). The use of chaîne opératoire allows the process of production to be related to the producer, and to the amount of knowledge required for a wide array of production processes (Wendrich, 2006, p. 269).

Lemonnier linked chaîne opératoire to Mauss' (1936) theories of ways of using the body to industrial manufacturing. Mauss considered that techniques were integral to the everyday reproduction of society. Any change in technique should also be viewed as offering insight into changes in social behaviours as the two are symbiotically linked (van der Leeuw, 2002, pp. 239-240). The study of the chaîne opératoire of production enables the various processes and techniques that the manufacturer has to follow in order to achieve the product to be identified and related to the various expressions of the importance of the self, and the social identity of the person within the society through technical means (Dobres, 1999, pp. 124-5). The use of chaîne opératoire has been focused often upon the lithics industry and reconstruction of lithic manufacturing techniques, both through analysis of artefacts and coupled with experimental reconstructions, but ignoring the function of the lithic tools themselves (Pelegrin, 1990). It has rarely been utilised in ethnographic research, but has mostly been restricted to lab work or experimental work where some technological features are measured but are rarely tested in ethnographic studies (Livingstone-Smith, 2000, pp. 21-22).

Similarly, a range of manufacturing processes for identified objects such as pottery can be deconstructed into technical processes, and this deconstruction has been undertaken by various authors (e.g. Franken & Kalsbeek, 1975; Shepard, 1968; van der Leeuw, 1976; van der Leeuw, Papousek, & Coudart, 1991). This process enables an understanding of the possible choices and variations that the manufacturer could have used at any given point. It provides a clearer picture of the differences and similarities amongst human populations, such as religious, political, symbolic and economic pressures in relation to the expression of identity, ethnicity, age, gender and social status. Technology can both reaffirm and contest relationships and traditions (Berg, 2007, p. 235). Some ethnographic studies have shown how the agency of gender can be deeply embedded in the material world of resources and power, with some (elite) individuals having control of the objects produced, the craftspeople and the technologies and therefore control of the status of both themselves and the craftspeople (Herbert, 1993). Technologies allow agents to construct social identities and power relations as well as producing utilitarian objects (Dobres, 1999, p. 129).

The need or the drive for technological change is often led by a social need for new technology (Hill, 1977; Plog, 1974). Technology itself is largely self-contained, as once it is established, it will continue exponentially, but it will not be the driving force for other technologies, as it is not a physical creative entity. Rather, it is other external elements derived from the social, political and economic elements of society that serve as triggers for technological change (Mackenzie & Wajcman, 1999, p. 3). A change in technology will be only one factor amongst many others. If a technology is biological and the physical effects are complex, so too must be the social effects. Roux (2003, p. 3) states that technology and society can be seen as a never-ending feedback cyclically and mutually dependent, each constantly adapting as the other changes.

Specific technological choices are pursued based on underlying social, physical, and functional factors that instigate technology. Cultural and technological choices expand upon the interactions between technology and society, an idea first postulated by Lévi-Strauss (1973) and further expanded by Lemonnier (1989; 2002) and van der Leeuw (1976; 2002; van der Leeuw, Papousek, & Coudart, 1991). Technological features are either maintained or amended depending on the previously established symbolic, religious, and social significances within wider networks of meaning. The addition of new technological techniques to the already established ones depends on the level of coherence between existing cultural perceptions of material elements and the ability to perform modifications in the manufacture and use process. However, technological choice presupposes that social groups have been able to make a choice by selecting one technological solution or practice over a series of other possible choices (Roux, 2003, p. 8). This is often not the case as the technology in a society is socially embedded within it and is often reinforced by elites. Even in a completely egalitarian society, craftspeople often do not have many technological choices and are limited by the resources and machinery available, and the traditions of craft production already developed (Pritchard & van der Leeuw, 1984, pp. 11-12; van der Leeuw, Papousek, & Coudart, 1991, p. 147). For example, the mould-made pottery produced by the potters of Michoacán, Mexico will answer, "*Such is the tradition*" when their production methods are questioned (van der Leeuw, Papousek, & Coudart, 1991, p. 156). Other methods of studying technological choices in terms of an ecological or perception action perspective have been postulated by Bril *et al.* (1998) wherein skills occur as a dynamic interaction between the task, environment and the subject (Suchman, 1987).

Technologies in themselves are neutral and therefore cannot actively shape a society, but the way that societies choose to use technology can be politically motivated to change the structure of the society (Winner, 1999, pp. 28-30). Technologies can be designed to open certain social conditions but close others, either consciously or unconsciously (e.g. a society which uses a moulding technique to form its pots will be restricted to the shapes which can be carved into moulds whereas coiling pots allows a wider variety of designs and shapes). The use of some technologies can be entirely politically motivated, whereas others are more compatible with some social relations. The view that technology just changes of its own accord or follows "science" cannot really be applied to preindustrial societies, as the concept of science was not often separated from religion and logical thought as is the case in some "modern" societies (Mayr, 1976). A contrasting idea promotes a passive attitude to technological change in that technology just happens of its own accord. In the modern world, technology is linked with applied science, with scientists discovering facts about reality and technology and applying these facts to produce useful things. The thinking being that technology shapes technology and that great inventions occur in a "eureka" moment by one single person at a specific moment in time, which must not always have been the case. New technologies do not emerge from a single moment of inspiration, but from pre-existing technologies and through a process of gradual change and considerable thought, often involving a variety of people working together to a common goal. Technologies emerge as part of a system; it is not just as a series of techniques but a complex variety of functions of techniques, as well as economics, organisational, political, and cultural aspects. In adopting a technology, a society may be taking on many of these other functions (Mackenzie & Wajcman, 1999, pp. 3-11).

Technology then, can be said to be the result of dynamic and complex processes emerging from properties of the constituting components and the interplay of all the components involved- the techniques, the environment and the subject (i.e. the technique is coiling, the environment the clay deposits and inclusions, and the subject the potter). Technology is an open system, the result of continuous interactions and exchanges from within the technological and social domains which are in turn transformed through these interactions (Roux, 2003, pp. 9-12). Technology in the ancient world was not necessarily shaped by science, so had to rely on the empirical observations and the experiences or skills of the craftsperson(s). The ancient artisans and engineers would have had to proceed through trial and error, utilising their knowledge of the natural world, and known techniques, and through experiment and observation new technologies could emerge (Forbes, 1958, pp. 5 & 42).

Learning a craft can take a long period of time, as the craftsperson must master hand-eye coordination, with the skills to perform the correct action at the correct point in the sequence and the patience and endurance to withstand the continuous repetition of movements to complete the desired product. This process can involve muscle strength, correct physical positioning and steady movements of the hands (Colbeck, 1982, pp. 19-20, 24-57). Secure knowledge of the materials and their properties is also required, which together with learning the techniques could take a lifetime to learn. This is perhaps why many craftsmen and women would learn from their parents or be specially apprenticed to a master craftsperson in order to be fully absorbed into the trade from an early age, since a craft can only truly be mastered through a tactile approach; reading about a craft rarely gives the student sufficient knowledge transfer (Wendrich, 2006, pp. 273-4). Through learning by observation and constant repetition of movements coupled with the knowledge of the properties of the raw materials that the craftsperson was working with, the craft would gradually become so ingrained that it would be as though it was second nature, even though a great variety of craftspeople have no knowledge of the underlying science behind their craft (Schiffer 1972; Schiffer & Skibo, 1987, p. 597).

A series of pre-determinates have to be in place before a technology can take root in a society. (1) There must be a weakness in the current technology, (2) a tolerance for "new" things in society, (3) the invention must be technologically possible, and (4) the new technology must obviously be of benefit to the society by being an improvement upon the established technology. These issues will be deconstructed and compared to the development of the invention of potter's wheel (see *Reasons for Inventing the Potter's Wheel*). The general process leading to an invention can be outlined as follows. Weaknesses within the existing technology should have been identified, and since no technology is perfect, some aspect can usually be improved. The society must have a tolerance for allowing new ideas to be tested. In the Near East, if the city-state, or in the case of Egypt, the state, is totalitarian and new ideas are viewed as a challenge to the structure of society, any new ideas are unlikely to be accepted. The invention must be technologically possible, given the current state of knowledge and skill at the time of the invention. Where city-states contain the same level of knowledge and technologies, then simultaneous inventions of a comparable technology can often occur. The vast majority of inventions are continuous to prior technology, otherwise they often cannot be used e.g. Leonardo da Vinci's idea of the helicopter or "aerial screw" was technologically far too advanced to be created in the 1480s (Gablehouse, 1969). In some cases, rather like Archimedes in his bath crying "Eureka!" a new idea can spring upon an inventor without warning; such is the capacity of the human brain. The new technology has to be an improvement for the society, but it may take time before it becomes reality or is universally accepted as an improvement to all sectors of that society.

By understanding the conditions necessary for the uptake of a new technology, it is hoped to understand why the Egyptians adopted the use of the potter's wheel. The previous section detailed the issues behind developing a new technology and how a society begins the process of inventing. In the next section, these theories will be more actively applied to the invention of the potter's wheel. Archaeologists have sometimes made assumptions relating to the reasons for its invention i.e. the mass-production of pottery, increased speed and the change from a "domestic" female sphere to a "workshop" male sphere. The next section will question these suppositions and endeavour to identify alternative reasons for the conception of the potter's wheel.

Reasons for Inventing the Potter's Wheel

Scholars e.g. Childe (1954, p. 204; Foster 1959a, p. 101; 1959b) have suggested that the potter's wheel was instigated for the mass-production and standardisation of pottery, or that it allows for a greater variety of forms and less drying time as the pot is thrown in one piece (D. E. Arnold, 1985, p. 208). When considering pottery production in Egypt, hand-building techniques were highly developed by early Naqada I times (c. 3600 B.C.), with skilled potters producing functional yet stylish pots e.g. decorated ware, black-topped red ware (see Figure 4.2) made through sequential slab construction and coiling (Vandiver & Lacovara, 1985, pp. 53-85). These pots were almost certainly made by specialist potters in Upper Egypt, probably at Hierakonpolis who then traded and sold their wares as far afield as the Levant and Nubia (Friedman, 1996, pp. 16-35). The Predynastic town of Hierakonpolis reached a population of between 5-10000 people by 3400 B.C. The town contained a central area with a temple dedicated to the god Horus of Nekhen, with various zones around it dedicated to different industries (Hoffman, 1982). Hierakonpolis seems to have been a major producer of pottery, beer, stone vases, mace heads, palettes and other commodities, which it exported throughout Upper Egypt and beyond (Wenke, 2009, p. 222).

Standardisation of pottery is often the most commonly cited reason for the advent of the potter's wheel (Bourriau, Nicholson, & Rose, 2000, p. 142). Standardisation is largely defined in the literature as a relative degree of homogeneity or reduction in varibility in artefact characteristics; in the case of pottery this signifies form, decoration and paste composition (D. E. Arnold,. 2000, p. 334). The identification of standardisation is usually determined by a comparison of two or more artefact assemblages with differing degrees of homogeneity (Blackman, Stein, & Vandiver, 1993, p. 61; Costin, 1991; Rice, 1981, p. 268). This implies that there must be some form of specialisation of pottery that is mass produced. Whereas, a more heretorgenerous assemblage indicates "household" production (Blackman, Stein, & Vandiver, 1993, p. 61). If such production was centralised, one would expect pottery production to be highly standardised and homogenous throughout Egypt, whereas localised

FIGURE 4.2: AN EXAMPLE OF PETRIE'S BLACK TOPPED WARE, UC9546 ©PETRIE MUSEUM OF EGYPTIAN ARCHAEOLOGY, UCL

FIGURE 4.3: AN EXAMPLE OF A BEVELLED RIM BOWL, ©ASHMOLEAN COLLECTION 1981.986. PHOTO: S. DOHERTY

production would be reflected in relatively heretogenerous assemblages (Sterling, 2004, p. 3).

However, even within the context of hand built pottery, the Egyptians already had specialists involved in the mass-production of pottery in a uniform style, but still beautifully made. These potters engaged in long distance trade by c.3500 B.C. They had no need of the potter's wheel to speed up production, as they had already designed the *bedjˤ* bread mould. This was the same in the Ancient Near East. Before the potter's wheel was in use, potters had already mastered the art of mass-production and standardised forms they worked out that if they used a mould then they could quickly and efficiently create the same ware as many times as they needed. From the early 4th millennium, the bevelled rim bowl (see Figure 4.3) was created in vast numbers in temples throughout the Near East. These were heavily tempered bowls designed for risen bread baking. They were shaped in either the ground, using a wooden mould or a premade and fired bowl (Nicholas, 1987, p. 60; McAdam & Mynors, 1988, p. 40). Since they occur in a variety of different sizes, they cannot be used as an example of standardisation of pottery, but can as an example of mass-production. These pots require very little skill on the part of the potter and

FIGURE 4.4: REPRESENTATION OF A SHRINE ON THE TOP OF THE URUK VASE. SHOWING FROM LEFT: MAN CARRYING BASKET OF OFFERINGS TO PRIESTESS IN FRONT OF SHRINE OR TEMPLE OF GODDESS IANNA. SHRINE IS REPRESENTED BY TWO REED BUNDLE STANDARDS, WITH STREAMERS WHICH ARE SYMBOLS OF THE GODDESS. IRAQ MUSEUM, IM19606, EXCAVATION NUMBER: W14873. CALCITE, C.3000 B.C. ORIGINAL HEIGHT C 105CM, UPPER DIAMETER 36CM. PHOTOS: HIRMER VERLAG

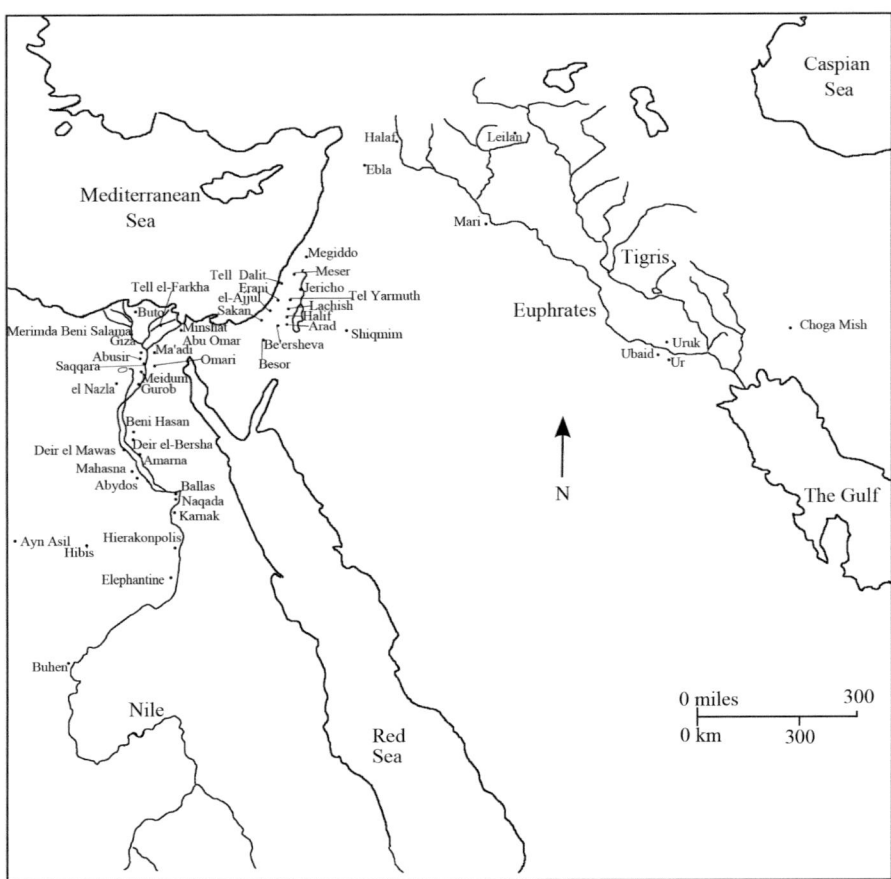

FIGURE 4.5: MAP OF THE NEAR EAST AND EGYPT, SHOWING KEYS SITES MENTIONED IN THE TEXT. S. DOHERTY

could have been made by most people (see Figure 4.3). They were fired at low temperatures 700°C (possible in a hearth) and so could serve both to raise the dough and to bake the bread inside it. Prior to use, they could be stacked and heated before adding the dough (Goulder, 2010, pp. 351-355).

The bevelled rim bowl may have had only a single use, as many have been found lying upside down, stacked and intact in or near temple precincts. Beale suggests (1978, p. 305) that the bevelled rim bowl was designed to allow the presentation of a token amount of a commodity as an offering to the gods or priest king at a temple, administrative centre or shrine similar (see Figure 4.4) to that depicted on the "Uruk Vase" (Basmachi, 1947).

The bevelled rim bowl was discarded once used e.g. at Chogha Mish individual pits were found to contain hundreds of bevelled rim bowls (Van Buren, 1952). However, as everyone could potentially make them, it could be that these bowls could have been made by everyone within the temple workshops, which would also explain the diversity in rim sizes. Therefore, the potters of the Near East did not need the potter's wheel to increase their production of pottery, they knew that the mould was the most effective means, as used by both South American potters and the pottery industries of the UK in the 17-20th centuries A.D.

Developing the technology of the potter's wheel

The technological situations in Egypt and the Near East will now be investigated, in order to ascertain whether they were capable of inventing the potter's wheel, given the resources available and the state of technical knowledge in the 4th millennium in these areas. These societies would have had to have a tolerance for such a new technology to be developed, before it would be accepted.

The situation in the Near East and Egypt (c.4000-2600 B.C.)

For this discussion, the geographical focus of the Near East comprises Mesopotamia (Iraq and Syria), Iran, and the Levant (southern sections of Lebanon and Syria, the Palestine Autonomous Authority, Israel, and Jordan). Some scholars add Egypt to this list, but for the purposes of this discussion Egypt is omitted from the countries of the Near East to aid comparison. Within this area, diverse ecological zones include: the Mediterranean coastal plain, widening from north to south; a central hilly zone between the coastal plain and the Rift Valley; the Rift Valley (including the Sea of Galilee, Jordan River, and Dead Sea); the Transjordanian plateau and escarpment; and to the east, the Eastern Desert extending into Iraq and Saudi Arabia. The Near East is therefore located just next to Egypt, and would have been one of its closest neighbours, (see Map at front of text and Figure 4.5).

During the period known as the Chalcolithic (4000-3000 B.C.) a great many specialised changes occurred throughout the southern Levant. Most notable were changes relating to mortuary and ritual practices, an increase in urban settlement patterns and the development of iconographic and symbolic expression. Rowan and Golden (2009, p. 2) suggest that the prime reason for this increase in specialisation was the introduction of agricultural intensive farming methods with bureaucratic officials organising land ownership and taxation. These underlying social, economic and political processes may have led to an increase in craft specialisation, through trade and long distance resource procurement, the desire for increased pottery production and ultimately the invention of the potter's wheel.

Evidence from both Egypt and the Levant in the 3rd millennium B.C. suggests that there is likely to have been contact between the two areas (see Chapter 5). There are similarities between Levantine and Egyptian pots from the 3rd millennium B.C. As at that time, some areas of Canaan were possibly colonies (Braun, 2003, p. 24; Gophna & Van den Brink, 2002, pp. 280-281). For example, Tell Erani (Brandl, 1989, pp. 357-388; 1992, p. 441), and Tell es-Sakan (Yekutieli, 2004, p. 171). The first use of wheel-made pottery seems to have occurred at some point in the Uruk period in Mesopotamia (4th millennium B.C.) (Simpson, 1997b, pp. 50-1). During this time, between the Ubaid and Early Uruk phases, the pottery changed from highly painted handmade wares to relatively plain and utilitarian wheel-made wares (see Figure 4.6).

These changes are not thought to have occurred when the Sumerians invaded Mesopotamia c.3000 B.C., but rather as an internal development of new technologies and innovative machines possibly instigated by the increase in metal and stone drilling production, which occurred shortly before (Kuhrt, 1995, p. 22). During the Late Uruk phase (c.3000-2900 B.C.) updraught kilns and clamp kilns[27] began to be used. A potter's quarter excavated at Ur contained circular kilns with shallow fire pits 35cm deep by 90cm across supporting perforated clay grates. Near this kiln, a ceramic disk wheel was found with dimensions of 75cm diameter, 5cm thick, and weighing 44kg (see Chapter 2, Figure 2.3 and Table 2.1). This suggests that the updraught kiln and the wheel could have developed simultaneously in Ur, and possibly throughout the Near East and Egypt.

The Development of the Near Eastern City States

Rather than being under the control of one ruler, the Near East developed a conglomeration of city-states, clustered near to the great rivers Euphrates and Tigris and around the coastal regions of the Mediterranean (see map). Each had a different governor or king, in some cases paying tribute to a larger polity e.g. Ur. The city-states of the Near East developed highly complex ownership and land management systems of irrigation and agriculture that instigated a food surplus for the first time since so many people were set to work the land by bureaucratic officials. This food and resources surplus allowed some individuals to give up farming and concentrate on crafts or administration within the central economic authority developing in the temples (Knapp, 1988, p. 39). The vast majority worked the land to supply the ever-demanding requirements of middlemen and courtiers. This led to an increasing trend towards monumental architecture and political organisation during the Uruk period (late 4th millennium B.C.) in turn widening the social divide and the demand for superior quality goods to furnish this new richer lifestyle for 1-2% of the populace. Trade and commerce was placed under the control of a few selected elite families who controlled the input and outgoings of trade and the production of luxury goods in order to increase their own wealth (Knapp, 1988, pp. 39-40).

Centralised control of the temples and the court officials

The control of the temples in the Near East seemed absolute. These temples e.g. of Ebla, Leilan, Ur, and Uruk (Stein & Blackman, 1993, p. 33) housed and managed the resources of the leading craftsmen who were producing high status goods e.g. precious metals, beaded jewellery (e.g. carnelian and lapis lazuli probably imported from Iran and Afghanistan). Land tenure was recorded and controlled, employment of workers was organised by the temple, and it also served as a granary. The temple became the centre of the organisation and distribution of goods to the surrounding city and its environs. It had a large

FIGURE 4.6: POLYCHROME HANDBUILT POTTERY © TRUSTEES OF THE BRITISH MUSEUM AN144655

---

[27] Unlike the updraught kiln, clamp kilns are a temporary structure in which the pots are stacked and baked in a pit underneath a bonfire.

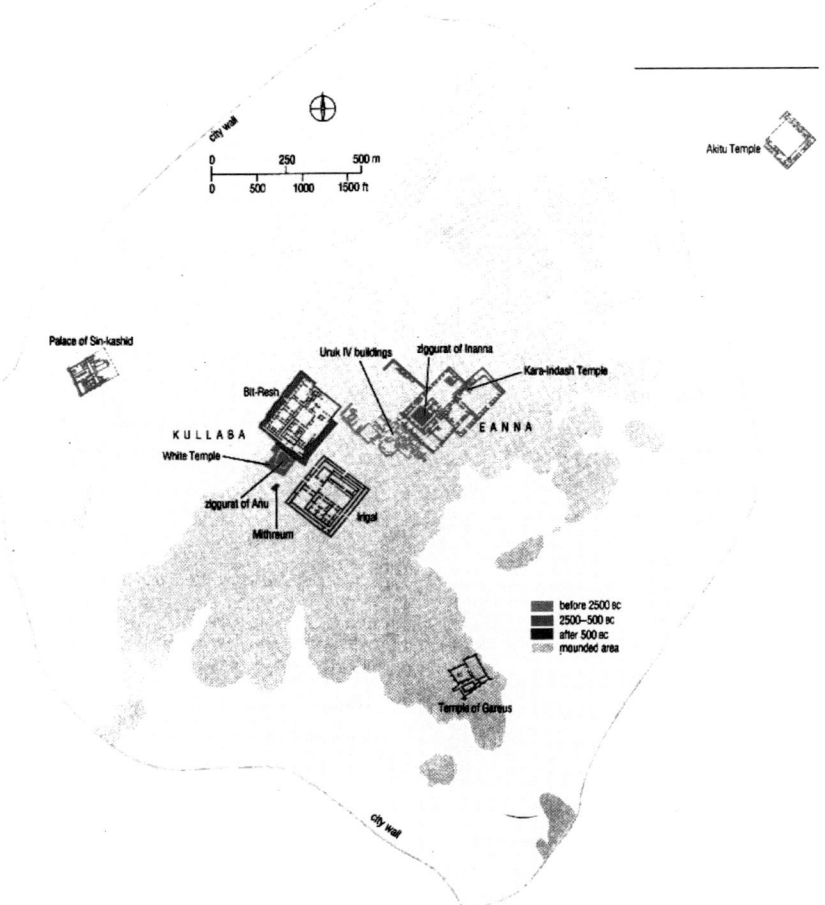

FIGURE 4.7: THE PLAN OF THE CITY OF URUK-WARKA, SHOWING THE TEMPLE COMPLEXES OF ITS PATRON GODS IN THE CENTRE OF THE CITY ON THE HIGHEST GROUND. WITHIN THESE PRECINCTS, CRAFT WORKERS' WORKSHOPS WERE LOCATED. ROAF 1990, PG 60, FIG 60B

influence on technological changes and developments as it controlled what the craftspeople were producing within its walls, and where and to whom those goods would be distributed. A hierarchy was created amongst the cities' inhabitants, with the more prestigious goods going to the elite families who could afford them (Knapp, 1988, p. 43).

The temples may have begun (during the early Ubaid times c.3500 B.C.) as large stockpiles for resources during times of famine or hardship, allowing them to maintain control and support the well being of the people by controlling the distribution of resources for the good of all the community (Stein, 1994, p. 44). Items such as food and textile processing equipment found in the temple suggest that the temple was also producing goods similar to that expected in domestic contexts (Pollock, 1999, pp. 87-88). Matthews (2003, p. 105) suggests that the inclusion of domestic activities in the temple may in fact represent them into ritual activities e.g. baking of holy bread and serving holy beer in particular pottery vessels.

The city of Uruk-Warka (for which the period has been named) appears to indicate these changes in its city plan. It was the largest city of the time (3500-2900 B.C.), excavated by Robert Adams and Hans Nissen in the 1960s-70s (Adams & Nissen, 1972) comprising some 250 hectares and possibly supporting a population of c. 25,000 people (Nissen, 1988, pp. 71-72). Uruk, rather like the later Greek city-states e.g. Athens and Corinth, had a raised acropolis-like area in the centre of the tell where the temple was located. The sacred areas were kept slightly apart from the secular, despite being the centre of economic life, and the surrounding city was divided up into neighbourhoods; residential, administrative (palaces), industrial and a cemetery. According to the epic of Gilgamesh, one-third of the city was temples, one-third houses and one-third gardens (Roaf, 1990, p. 60). There were no distinct rich and poor residential areas, but the temples, while being the administrative hub, were fully detached from the rest of the city (see Figure 4.7). The neighbourhoods were also divided by streets, canals, and water channels. Rather like modern day Venice, the canals were part of a large regional system with cities established along the banks; this allowed access to a series of different markets along the Tigris and Euphrates rivers (Gates, 2003, p. 32).

Most importantly, there was a high tolerance and openness for new ideas in the city-states of the Ancient Near East, albeit strongly regulated. This would have had a large influence on any technological changes and developments as well as controlling what the craftspeople were producing within the city walls. If the temple decided not to fund a new idea, then the idea probably was never fostered nor developed and the craftspeople would be put to work on something preferred. As the temple personnel seem to have decided that the potter's wheel was to be developed, they were able to instruct the stonemasons under their jurisdiction to start making the pierced stones needed to create the wheel bearings.

The Situation in Egypt c.3500-3100 B.C.

At some point in Egypt's past, about 3500 B.C., two or three cultural entities existed with differential access to wealth, power and prestige (Wenke, 2009, p. 205). One society occurred in Upper Egypt in the towns of Hierakonpolis, Naqada and Abydos (Kaiser, 1985, pp. 61-87) and a rival system was found in Lower Egypt in the areas of Ma'adi and Buto (Kemp, 2006, pp. 31-35, fig 13). The process of the development of the early Egyptian state formation must have been at least in part driven by the Upper Egyptian rulers' desire to obtain and control the prestige goods networks and to enable access to trade with the lands to the north and the south; such as the basalt quarries in Gebel Qatrani, the gold from Nubia and for commodities from Syria-Palestine (Ekholm & Friedman, 1982, pp. 87-109). These would have been actively sought by the elites for the conspicuous consumption of rare and valuable materials (Wilkinson, 2001, p. 113) that set the upper classes apart from the other regular Egyptians. By being active consumers of these goods the elites could have a visual way of displaying their power (Peregrine, 1991).

Lower Egypt's close contact with the Near East perhaps initially gave it the edge over Upper Egypt (Braun, 2003), as it was able to take advantage of new technologies such as stone vessel making, which is attributed by Woolley (1955) to have begun in Palestine (although no drilling devices have been found either in Egypt or the Near East to prove or refute this claim). However, perhaps the location of the Upper Egyptian cultures at This (near Abydos, possibly located at the modern Egyptian town of Girga see map) and Naqada allowed it in the end to dominate the more disparate communities and villages of Ma'adi and Buto in the delta. Both these areas had good access to neighbouring societies in the Sudan and the Levant, and so could trade and exchange goods and ideas.

However it happened, towards the end of the Naqada II period (c.3300 B.C.) a substantial change in the way of life occurred in Lower Egypt, possibly indicative of Upper Egyptian supremacy; through competition, diplomacy or even violence. The changes can be noted in the buildings, for example, in Buto mud brick architecture began to be utilised for the first time (Faltings, 1998b, pp. 365-375). Prior to this wooden posts and wattle and daub style housing were the norm, such as that recorded at el Omari (Debono & Mortensen, 1990). Alongside this, pottery of the Naqada traditions (e.g. black topped and red wares) was introduced and changes in the type of flint tools occurred (van den Brink, 1989). At Minshat Abu Omar, a Naqada style cemetery dating to the late Naqada II period has been located (Kroeper & Wildung, 1994). By Dynastic Times (c.3100 B.C.), the cultures of the Nile Valley from Elephantine to the Delta became homogeneous through the expansion of the Naqada cultures spreading northwards from the main southern towns of Naqada and Hierakonpolis, (and later the town of This), compelling their own material culture, technology and no doubt language upon the north (Kemp, 2006, p. 89). However, this could also be explained as a difference in dialects[28] rather than a distinct language (Assman, 1996, p. 29).

During the Naqada III and Dynasty 0 periods (c.3200-3100 B.C.), Egypt, unlike the Near East, became unified into a single nation state under one ruler, the Pharaoh. Once unification was accomplished, possibly by Pharaoh Narmer or Scorpion, over 1000km from Buto in the Nile Delta to the first cataract at Aswan was under the jurisdiction of the Pharaoh. From his capital in Memphis, this pharaoh quickly established strong administrative and bureaucratic control over the region through taxation, royal monopolies of resources, expeditions to foreign lands and military campaigns using conscripted soldiers (Bard, 2000, pp. 67-8, 87-8; Wenke, 2009, p. 189). Most likely the introduction of a common language and a writing system aided this process (Baines, 1983). The power of the Pharaoh, the royal family and his court was absolute. The construction of large tombs utilising a corvée workforce and royal mortuary cults at Abydos testify to this ideology (Kemp, 1966; O'Connor, 2011). The use of writing and military imagery upon previously everyday objects such as cosmetic palettes e.g. the Narmer Palette, mace heads e.g. king Scorpion, ceramics e.g. white painted wares and knife handles e.g. the Gebel el-Arak knife cemented the king's authority and recorded his deeds for posterity. Recording the names of kings became significant, for example, the basalt Palermo stone demonstrates the divine right of the king to rule from the beginning of time (Assman, 1996, p. 38). Such cultural markers meant that a select amount of elite males were in control over the rest of the nation. Most of the textual, artistic and archaeological sources are derived from this small group of men, and much of the production and consumption of aesthetic items were solely for their benefit (Baines & Yoffee, 1998, p. 235).

The technology available in The Near East and Egypt c.3500 B.C.

The people living in the Near East in 4000-3500 B.C., when the potter's wheel was invented (see Chapters 2 and 3), used relatively few machines in their industries. To increase production, they simply multiplied up the

---

[28] A text reflecting the north-south accent barrier notes: "there is no one conversant with foreign tongues who could explain it. It is like the conversation of an inhabitant of the Delta with a man from Elephantine" Papyrus Anastasi I 28, 6. Such a difference can still be heard today if one compares a Cairene accent to a Saeedi (Upper Egyptian) one.

numbers of workers (Shaw, 2004). Consequently, if a new machine were to be introduced to such a non-industrial economy, there had to be a good reason for doing so.

Copper working, stone carving and drilling (Stocks, 2003) were all quite technologically advanced during the 4th Millennium B.C. The ability to work and smelt copper meant that specialised craftsmen must have had knowledge of creating high temperatures in excess of 1100°C in furnaces using bellows or blowpipes (Davey, 1979; Nibbi, 1987), and probably of making and using charcoal (Ogden, 2000, pp. 149-155). Therefore, ancient craftsmen were aware of how to control and maintain the temperatures of kilns and were likely to be working in workshops, in association with temples or large estate owners occupying high positions at court.

Just as in the Near East, the temples and the court of Egypt were deeply involved with the craftsmen's lives and work. By ensuring that the workshops were attached to the temples and royal estates, temple personnel and courtiers were able to control the resources available to the craftsmen and so would have been able to govern what the craftsmen were making. When the craft and the products resulting from the craft were so embedded within the temple domain, the items produced by the temple workshops would have been imbued with a special ritual significance by the temple priests. For example, if a craftsman makes a wooden statue within the temple, it was not yet fit to serve its ritual purpose until a priest had performed the rite of opening the eyes and mouth. The priest was therefore infusing the statue to have magical properties enabling it to be a suitable resting place for the soul of a deceased person (Forbes, 1958, pp. 40-1). The statue can therefore fulfil its religious function either in the temple itself or in the funerary sphere in tomb or chapel. The hardwood used to make the statue would have been specially selected and brought to the workshop in the knowledge of its ultimate use. Such wood stocks would have been highly regulated by the state, and its redistribution to craftspeople and their location within the temple makes economic sense when controlling production of such pre-eminent commodities.

Similar situations may have been taking place in the funerary customs of the Old Kingdom Egyptians during burial of a deceased person and subsequent cultic activities after the funeral. Vessels relating to the funeral e.g. cooking pots, bread moulds and luxury tableware (i.e. Meidum bowls, see Chapter 7) were possibly made by potters working in the funerary necropolis (Rzeuska, 2006b, pp. 353-357). This is suggested by the tomb of Ptahshepses (Krejčí, 2009, p. 145; Vachala, 2004b) where the potter is recorded as *The Potter of the House of Eternity* cemetery/mortuary estate Weri (Senussi, 2006, pp. 329-30; Warden, 2010, pp. 185, note 4) and in the potter's workshop discovered within the pyramid complex of Khentkaus at Abu Sir (Verner, 1992, pp. 50-5). Such examples could indicate that the potter's workshop was state organised. From the archives of Neferirkare Kakai (Posener-Kriéger, 1976, pp. 631-634) comes the suggestion that the products destined for the sun temple of Neferirkare Kakai, were derived from agricultural domains established by the king from different parts of Egypt to support the construction and maintenance of the funerary cults (Vymazalová, 2011, p. 296). Presumably, these products were also sustaining the potters at Abu Sir who were manufacturing the pottery intended for the cult.

Basalt and hard stone vessel production

In this section, the importance of basalt and its use in the construction of potter's wheels will be considered as a case study for the production of a state-commissioned craft. Basalt is a stone that is difficult to source, procure and work. It was used initially for the production of stone vessels, but as has been noted in Chapter 2, it was often the stone of choice for producing the bearings of potter's wheels. Given that the Egyptians could have chosen softer stones (and it seems in later times they often selected limestone, see Chapter 2, Tables 2.1 and 2.2), the reasons why basalt was the first choice stone will be considered here.

From the Predynastic Period, stone workers were increasingly using tools such as drills to work stone to form vessels, mortars, statues and funerary *stele*. As the potter's wheel bearings were often made of rounded and pierced basalt or diorite, potters relied upon the expertise of stone craftsmen to carve their wheels for them. Copper was perhaps formed into thinly beaten tubes and placed around a hollow wooden tube and used as a drill. These were further strengthened by a strong stick upon which one or two stone weights were attached by means of netting or bags with an inclined or tapering handle (see Figure 4.8). This sort of drill is otherwise known as the "Twist Reverse Twist Drill" (TRTD) a self-explanatory term to explain the movement involved in the drilling process (Stocks, 2003, pp. 142-3). The Drill is twisted 180° and then turned back to its original position. Although examples of the twist reverse twist drill have so far not been found archaeologically, they have been represented pictorially e.g. in the Tomb of Ty (Steindorff, 1913) or they can be seen in tomb models in conjunction with stone tool making e.g. Model of Inpuemhet and Usermut, Cairo Museum (see Chapter 3).

The bow drill may have been drilled rather like a fire starter, utilising centrifugal force just like the potter's wheel. This would have created a tapered hole, which the stone vessel makers may not have always wanted, but which would have been perfect for creating the well in the socket of a potter's wheel or piercing both bearings as seen in the earliest examples found in the Near East (see Chapter 2, Table 2.1).

Some physical evidence for the early use for the TWRD comes from the Naqada II temple of Hierakonpolis. From the temple revetment, a door socket made of dark quartzite was found with a jamb standing in it and a pivot hole for the jamb on the top (see Figure 4.10 for findspot). It was left rough on most of the sides so that it could be built

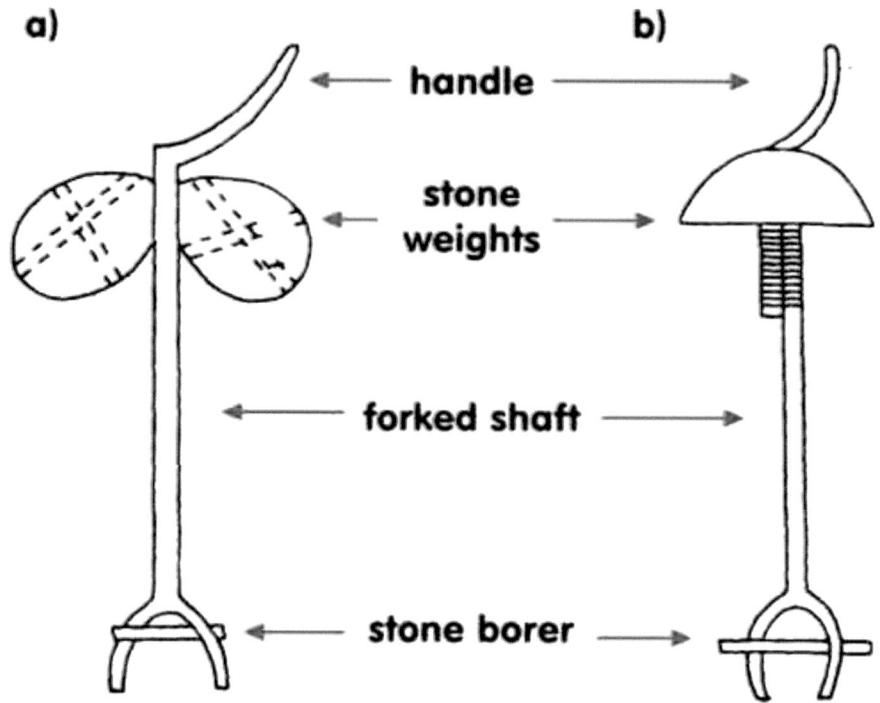

FIGURE 4.8: THE TWIST REVERSE TWIST DRILL. A) OLD KINGDOM EXAMPLE WITH 2 STONE WEIGHTS, GARDINER'S U25 DETERMINATIVE B) THE NEW KINGDOM VARIETY WITH A SINGLE LIMESTONE WEIGHT. THESE WOULD HAVE EITHER A FORKED SHAFT ATTACHMENT AS SHOWN, OR A HOLLOW BORER WITH COPPER TUBE ATTACHMENT. AFTER STOCKS 1993, P. 598

FIGURE 4.9: DOOR SOCKET MADE OF QUARTZITE FOUND NEAR TO THE TEMPLE REVETMENT AT HIERAKONPOLIS, AND ABOUT 10M AWAY FROM THE MAIN DEPOSIT WITHIN THE TEMPLE ENCLOSURE. NOTE THE SOCKET FOR THE PIVOT OF THE DOORJAMB AND THE HUMAN HEAD TO LEFT. AFTER QUIBELL J. E., 1900, PLATE III

into the surrounding masonry. From one corner, a human head has been carved, presumably an early example of enemies being crushed underfoot by the victorious Pharaoh, which was a more popular trend in later Dynastic times e.g. the Ramesses III palace at Medinet Habu (see Figure 4.9 and compare with Figure 4.12, the granite door jamb at mortuary temple of Niuserre). The key point is that it demonstrates that the Egyptians were capable of making a socket joint for temple architecture from at least the Naqada II period (c.3450 B.C.) in a manner very similar to the socket and pivot joint system required for the construction of a potters' wheel, no doubt using the TRTD described earlier. If the Egyptians were not actually manufacturing potters' wheels at this time, at least they had the technical knowledge and the appropriate tools to be able to do so.

The sources of basalt in Egypt that the TWRD could have been used to drill are relatively rare, although basalt vessels dating to the Predynastic period are known from sites throughout Egypt e.g. Saqqara (El-Khouli, 1978, p. 789), Abydos (Petrie, 1977). The only known ancient basalt quarry is that of Gebel Qatrani in the Fayoum (Harell & Brown, 1995). It is often assumed that this area was the main source for basalt vessels (Lucas & Harris, 1962, p. 62) although subsequent modern day quarrying may have removed any ancient traces e.g. at Abu Roash and Abu Zabaal. The relatively small size of the vessels may indicate that boulders or broken off outcrops of basalt were used to carve the pieces, or that they might have been picked up and transported to be worked elsewhere (Mallory-Greenough, Greenough, & Owen, 1999, p. 1270; Rizkana & Seeher, 1988). Workshops have as yet not been found next to the areas of basalt sources; namely, the Haddadin sequence west of Cairo, East Cairo (Cairo-Suez Road), Bahariya Oasis, Middle Egypt (Zarouk and Minya) and Southern Egypt (mostly alkaline basalts near the Red Sea- the Natash volcanic field, see Figure 4.11) as identified by Klemm and Klemm (1993, p. 315) in their quarry survey of Egypt.

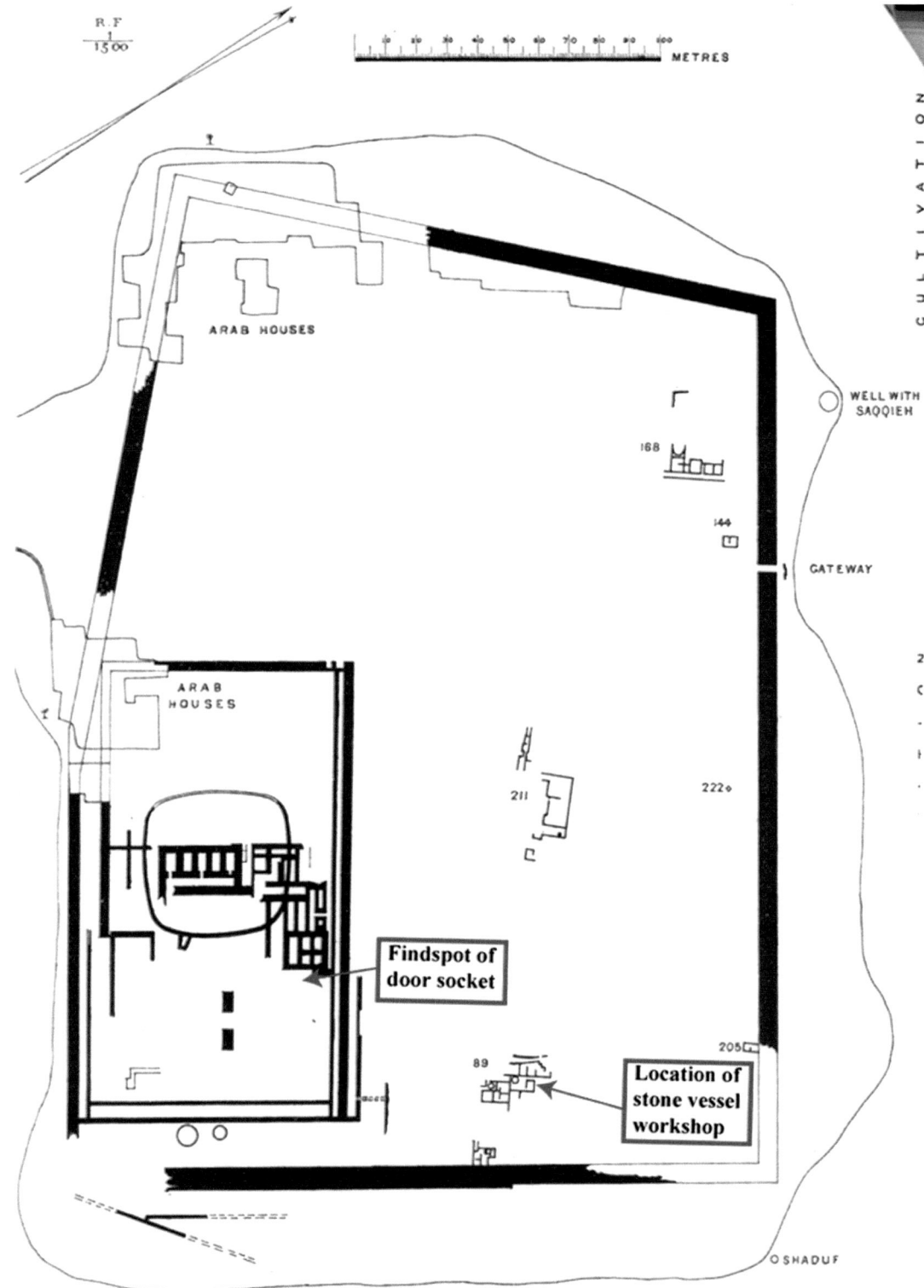

FIGURE 4.10: MAP OF THE TOWN OF HIERAKONPOLIS, WITH THE FINDSPOT OF THE QUARTZITE DOORSOCKET AND STONE GRINDERS' WORKSHOP INDICATED AFTER:QUIBELL 1900, PLATE LXXIII

The basaltic flows in Egypt are mineralogically quite homogeneous, so it is difficult to pinpoint an exact source for the quarries, although the one known ancient quarry at Gebel Qatrani seems the most likely source (Mallory-Greenough, Greenough, & Owen, 2000, pp. 323-326). The stone vessels may have been both quarried and carved in the north, and then transported south. Many stone vessels have been found in the north during the Predynastic period (Hayes, 1953, p. 23) and many unfinished ones have been found in Saqqara (El-Khouli, 1978, p. 789), which may be consistent with the first smelting and casting of copper at Ma'adi during the Naqada II (Amer, 1933; Stocks, 2003, p. 12) and hence the production of drills capable of drilling basalt. At the very least, if the source of this basalt

FIGURE 4.11: MAP OF EGYPT SHOWING BASALT OUTCROPS (SOLID BLACK) AND AREAS CONTAINING TRIASSIC TO TERTIARY FELSIC AND/OR MAFIC FLOWS, SILLS, DYKES AND PLUGS. (MALLORY-GREENOUGH ET AL. 1999 PG 1263)

was the Fayoum, then it was being transported all across Egypt from the Late Neolithic to the Old Kingdom periods solely for the production of basalt vessels. Gebel Qatrani seems a likely spot for an ancient quarry as it consists of an exposed outcrop of basalt on the uppermost ridge of the site which weathers continuously and regularly detaches blocks which collect on the slopes below (Aston, 1994, p. 20). If the raw basalt was being transported from the Fayoum to the workshops of Hierakonpolis, it was being moved some 600km. The administrative work relating to this transportation would have involved organising quarrymen, ships, personnel and soldiers for protection in order to procure something specifically for the elite funerary furniture or temple deposits.

Basalt became the stone of choice for Pharaonic sarcophagi and for the mortuary temple floors of the pyramid builders at Giza and Abu Sir (see Figure 4.12). Its dark black colour seems reminiscent of Geb and the black land synonymous with the temple as a microcosm of the world (Hoffmeir, 1993, pp. 117-120). The hieroglyphic for pyramid *mr* means "stairway to heaven" and so was the physical manifestation of the pharaoh travelling up to the afterlife. The temple was where the relevant rituals were undertaken. The roof of the temple was usually decorated with duat stars e.g. Unas causeway at Saqqara, Sahure's upper temple at Abu Sir which may also be connected to Nut, goddess of the sky and heavens. The temple can be viewed as a microcosm of the world, black earth below (Geb), sky above (Nut), air in between (Shu) (Finnestad, 1985, pp. 12-13; Reymond, 1969; Kemp & Rose, 1991, p. 103).

Stone vessels were one of the most common items of funerary equipment used by the Ancient Egyptians. Their earliest recorded use of stone vessels is in the Merimde Beni Salame culture in Lower Egypt (4800-4200 B.C.). Large-scale stone vessel manufacturing was established during the Naqada II (c.3600-3200 B.C.) and II/Dynasty 0 (3200-3020 B.C.) periods. Even from the earliest times, specialised craft and technologies were in the forefront in the development of the Egyptian state. Although Egyptian stone vessels of the Early Dynasties are most ubiquitously made from calcite (also known as alabaster), e.g. several thousand stone vessels were discovered in the step pyramid of King Djoser, (although many had come from the robbed tombs of earlier kings), the production of stone vessels for funerary and temple offerings became increasingly

FIGURE 4.12: NIUSERRE UPPER TEMPLE, ABU SIR 5TH DYNASTY. BASALT BLOCKS IN SITU IN TEMPLE FLOOR, WITH OTHER BLOCKS OF BASALT AND GRANITE LYING ABOUT. NOTE THE REMAINS OF SOCKET JOINT IN THE CIRCULAR GRANITE BLOCK IN CENTRE OF THE PICTURE, POSSIBLY INDICATING A GRANITE COLUMN, AND USE OF THE TRTD AND THE EGYPTIANS' ABILITY TO MAKE SOCKET JOINTS. PHOTO: S. DOHERTY

important. So much so, that the use of alabaster vessels was included in the standard Middle Kingdom offering formulae *htp di nsw*, placed on all offering *stelae* and entrances to tombs (see Chapter 7).

By the Naqada II period (3600-3200 B.C.) there was a rapid expansion in the volume of vessels being produced, perhaps suggesting that a faster and more reliable method of production was introduced. This was particularly in the case of Hierakonpolis, where stone vase workshops have been located less than 50m away from the main temple of Horus of Nekhen (Kemp, 2006, pp. 196, fig 68; Quibell & Green, 1900, p. plate LXXIII, see Figure 4.10). Another workshop dating to the Old Kingdom has recently been located near Sheikh Said (Vereecken, 2011); suggesting that the temple personnel were regulating the vessel production. This conveniently seems to coincide with the first use of smelted and cast copper in the Predynastic town of Ma'adi, located south of the apex of the Delta (Amer, 1933; 1936). Ma'adi may have close intercultural and economic contacts with southern Palestine, as both cultures seem to have replaced their polished stone axes with copper ones almost simultaneously (Midant-Reynes, 2000, pp. 58-59).

Beginnings of the Use of the Potter's Wheel in the Near East

Most excavated potter's workshops in the Near East (see Chapter 2) with provenanced potter's wheel bearings in the Bronze Age sites occurred near to shrine or temple areas e.g. at Megiddo (Engberg & Shipton, 1934, p. 40; Loud, 1948, pp. 268, fig 13, pl 268:1; Wood, 1990, pp. 99, fig 1:1) and Hazor (Wood, 1990, pp. 16, 99, fig 1:8; Yadin, 1958; 1960), in a cave e.g. Tel Halif (Dessel, 2009, pp. 20-22, fig 7; Jacobs & Borowski, 1993) and Lachish (Magrill & Middleton, 1997, pp. 68-9,72, fig 6a; Tuffnell, 1958, pp. 291-3, pl 49:12-13), in palaces e.g. Tel Yarmuth (Roux & de Miroschedji, 2009, pp. 161, fig 5) or Tel Dalit (Gophna, 1996, pp. 112-113, 144-5; Pelta, 1996, pp. 171-185, fig 1 & 2) or in designated potter's quarters as in the case of Ur (Simpson, 1997b, pp. 50, fig 1). The location of potter's workshops therefore was important for both the potter and the elite sponsor. The elite sponsor evidently required the potter to have his workshop located close to administrative areas, so that his production could be controlled, but not so close for the kiln smoke to adversely affect the area.

The pottery produced on the potter's wheel was not initially utilised to mass-produce standardised pottery for economic gain. Rather than being an instigation of standardisation, the first use of the potter's wheel to produce pottery may be a key candidate of elite-driven technology for ritual purposes. The pierced wheel bearings as discussed above were made using basalt; a prestigious stone previously restricted to the production of religious statuary, and now it is used for an industrial purpose as part of an ancient machine. When the potters of the Near East invented the potter's wheel, (or rather when their elite sponsors allowed them to initiate its use) they had not utilised it to its full potential for throwing. Rather, they used it as a finishing stand for coil made pots. Some of

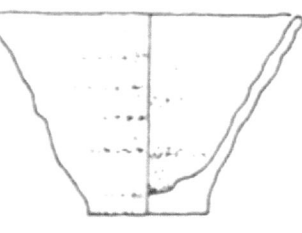

FIGURE 4.13: EXAMPLES OF V-SHAPED BOWLS, MADE BY ARRANGING COILS OF CLAY AND THEN THINNED AND SHAPED ON THE POTTER'S WHEEL. LEFT AND CENTRE: BM 125942; 1937, 1211.224 FROM TELL BRAK ©TRUSTEES OF THE BRITISH MUSEUM MIDDLE PHOTO: S. DOHERTY. RIGHT: PROFILE OF AN EXAMPLE THE V-RIM BOWL AFTER: ADAMS & NISSEN 1972, PG 309, FIG 6.G

the earliest vessels shaped on the wheel such as conical V-rimmed bowls (contemporaries of the mass-produced bevelled rim bowls mentioned earlier), were thinned and shaped using these large basalt wheel bearings. By using the potter's wheel to finish off the coil-made pots, the potters were actually adding to the amount of time that it took them to make their vessels, perhaps adding to its prestige and its importance to the elites as their use was purely ceremonial (Roux, 2009, p. 195).

The V-shaped bowls are shaped like an inverted and truncated V, have flat bases (the only bowls to have such at this date) and the walls form an angle of c135° from the base. The walls are very thin and the rim is simple (Dessel, 2009, p. 96). The type of V-shaped bowl found in the northern Negev and southern Shephelah e.g. Abu Hamid and Abu Matar in the Chalcolithic and Early Bronze Age I were wheel shaped, but not all were so (3500-3100 B.C.). The examples produced using a fine clay and little tempered sometimes had a red or white band painted around the rim, often a good indicator for the ceramicist that they were wheel fashioned (see Figure 4.13).

The V-shaped bowl (see Figure 4.13) occurred almost simultaneously, although not in large numbers, all across the Southern Levant in archaeological sites with the same cultural horizons, in sites such as Abu Hamid, Beer-sheva, En Gedi, and Halif (Commenge-Pellerin, 1987; 1990; Dessel, 2009; Perrot & Ladiray, 1980; Ussishkin, 1980). These bowls were at least partially formed on a wheel, first coiled and then drawn up using some form of rotary motion, which Courty and Roux (1995; Roux & Courty, 1998, p. 747) termed "wheel fashioning" or "shaping," but not through centrifugal force (which has resulted in the potter's wheel being falsely called a turntable or "*tournette*" when centrifugal force is not induced).

The use of centrifugal force is a key fundamental change in the techniques of pot construction. Prior to this, pots may be formed on a rigid support, such as a mat which the potter would rotate around while they were building and shaping their pot. For V-shaped bowls, the primary method for shaping the pot was through coil, pinch or slab, and the potter's wheel is used as an aide so that the potter can stay in one place rather than have to move around the pot while forming it. The potter's wheel in the Near East at this period was not rotated sufficiently fast enough for centrifugal force to be achieved, as evidenced by the marks on the pottery (see Chapter 6 & Figure 4.14). This is similar to the "banding wheel" or "whirlers" used by modern potters. Therefore, it could be argued that although the Mesopotamians invented the potter's wheel, they did not utilise it to achieve its full potential for throwing pots until much later.

Roux and Corbetta (1989) suggest that the use of the wheel for shaping pots and the use of the wheel for throwing pots represent two discrete evolutionary events that are not linked to the type of the wheel used. In their study of the potters of New Delhi, they found that in comparison to wheel-throwing, the coiling technique requires a much smaller investment of time to learn. Wheel-throwing involves a longer period of apprenticeship (10-20 years) before the potter is proficient, whereas wheel-coiling training takes only one year (Roux & Corbetta, 1989, p. 69). This suggests that wheel-throwing requires a considerable amount of effort on the part of the potter. If they utilised the wheel solely for shaping and positioning a coiled pot, the techniques involved do not greatly differ, whereas to produce a wholly thrown pot the potter needed to develop entirely new specialised perceptual motor skills (Gelbert, 1997, pp. 2-23; Roux & Courty, 1998, p. 748). In experiments making 63 V-shaped bowls, Courty and Roux (1995, pp. 17-50; Roux & Courty, 1998, p. 750) noted that when rotating the wheel whilst adding the coils and finishing the rim, the act of producing pottery becomes more mechanised, and therefore speed of production increased. However, it does not seem to increase speed of production if just using the wheel to finish rims. It seems that the instigation of the potter's wheel was not reduction in manufacturing time or mass production, but a social representation of wheel shaped ceramics and wheel fashioning methods that would have acted as symbols of urban elite identity (Roux & Courty, 1998, p. 761).

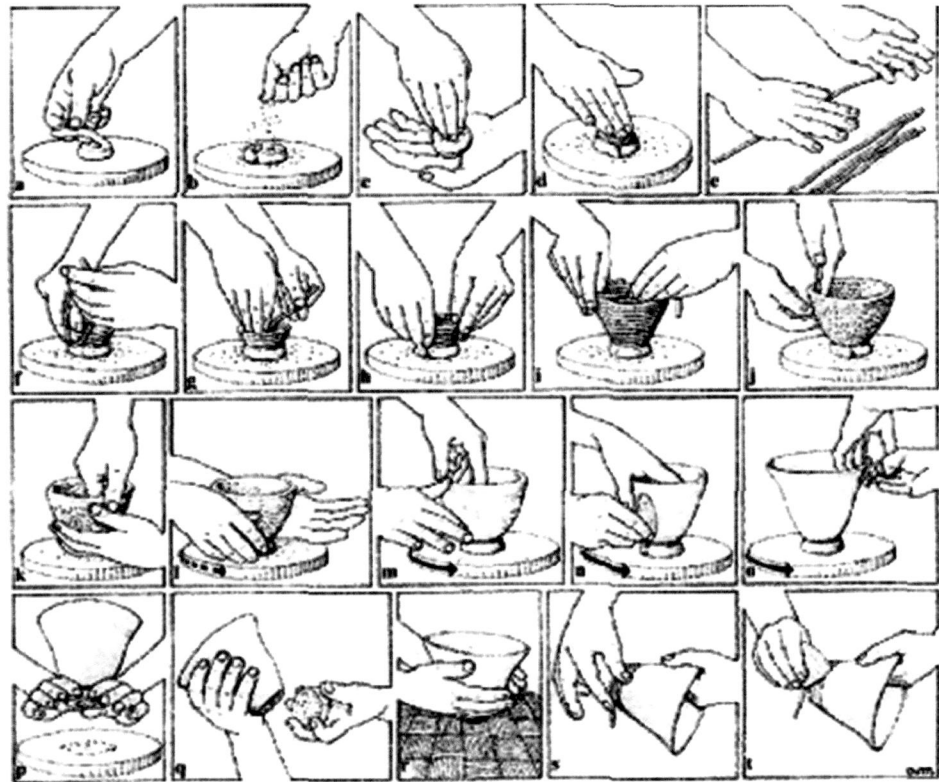

FIGURE 4.14: THE CHAINE OPERATOIRE OF THE V-RIMMED BOWL. (1) IT IS BUILT UP WITH COILS (LETTER A-I), UPON THE POTTER'S WHEEL. (2) THE WHEEL IS SPUN AND THE COILED POT IS THINNED AND SHAPED (J-O). (3) THE POT IS CUT OFF THE WHEEL AND THE BASE REMOVED (P-Q) (4) IT IS PLACED ON A MAT TO DRY. (5) FINISHING TOUCHES ARE ADDED AND THE POT IS SMOOTHED. ROUX & COURTY, 1997, FIG 1

The majority of V-shaped bowls are found in association with mortuary and cult centres such as Mezad Aluf, Gilat and En Gedi in ancient Israel during the Chalcolithic Period. The small amount of early wheel-shaped pots found within assemblages such as the temples at Abu Hamid, Abu Matar, Beer-sheva and Safadi (Commenge-Pellerin, 1987) suggest that production levels of wheel-made material were not very high, indicating that mass production was not a priority. V-shaped bowls have been found in a variety of contexts, in settlement sites, funerary and ritual. In the cemeteries of Shiqmim in the Northern Negev region of Israel, most individuals seem to have been buried with a V-shaped bowl; some cist structures in the cemetery without a burial contained a V-shaped bowl, perhaps indicating ritual activity (Levy & Alon, 1985, pp. 80-1, fig 3.3; Levy & Holl, 1988). They were also found in the sanctuaries of En Gedi and Gilat where they make up much of the ceramic assemblage (Goren, 1995; Ussishkin, 1980). These bowls might have been produced specially for ritualistic purposes by specialist itinerant craftspeople perhaps for the elite members for use in funerary and religious contexts (Roux, 2003, p. 22). Alternatively, the V-shaped bowls could have been especially imported from the Negev regions for the use of the local elite population. The presence of the V-shaped bowls along Abu Hamid and Beer-sheva nahal areas suggest a large community of people trading and perhaps sharing similar political and religious beliefs (Levy & Holl, 1988).

The potters themselves could perhaps have been part of a specialist emerging craftsman class, perhaps of low status and noted in later times in texts e.g. Satire of Trades. As Wendrich (1999, pp. 391-393) in her ethno-archaeological studies of basket makers noted, the skill of the basket makers was not linked to their professional status, but with the speed of their production. The potters could be supplying a need for new techniques and stylistic methods of producing pottery to make them inherently different from their handmade counterparts. The specialist techniques of wheel-throwing being important, rather than the decoration making them stand out. The demand for new vessels of ritual or funerary value by potters attached to the elites could indicate the broader changes soon to emerge within the political and religious fabric of the later 4th Millennium B.C. (Roux, 2003, p. 23). As discussed previously, generally the innovations in pottery production are derived from adapting tools and techniques in response to new social and economic conditions rather than the reverse, in particular when pottery production techniques are well-established (Miller, 2009, p. 188). However, with the invention of the potter's wheel, the connection between full time specialists and the switch from female potters making pots by hand and male potters utilising the potter's wheel and the beginnings of mass-production needs to be addressed in the next chapter.

Summary

The invention of the potter's wheel is likely to have been a cumulative process developed over time in the city state workshops of the Near East. It appears from the available evidence, that the updraught kiln, potter's wheel, and workshop developed almost simultaneously across the Near East as suitable social and economic conditions were in place in order to foster its use. The chaîne opératoire model can be applied to understand the various processes and techniques involved that the manufacturer has to follow in order to achieve the finished product e.g. the V-rim bowl. The notion that craftspeople often do not have many technological choices and are limited by the resources and machinery available and by craft production traditions already available can only be supported until a change in technology can be viewed as beneficial for the society. This could be argued for the development of the potter's wheel, as it met the elite members of society's new requirements for their funerary and ritual pottery needs. This has disproved the most commonly held assumptions regarding the advent of potter's wheel, that it was created for standardisation and mass-production of vessels. This does not seem to be the case. It was initially created to furnish the elites with ritual and funerary vessels, elaborately manufactured as part of their increased luxury lifestyles. The use of potter's wheels was seemingly strongly controlled by elite temple personnel, who would also have guarded who would have had access to the vessels being produced by the potters. The Egyptians would have been able to easily adapt to the Near Eastern model. Egypt was unified under one leader, the Pharaoh, whose court would have controlled its craftspeoples' access to resources by attaching artisans' workshops to temples and estates. This would have meant that any new machinery introduced to Egypt would have needed the financial backing of an elite sponsor in order for it to be implemented.

In this section, the reasons for inventing a technology and the significance of technological precursors for the potter's wheel in both Egypt and the Near East have been considered. Such evidence points to the premise that the Egyptians would have had the tools and the technology available to construct a set of potter's wheel bearings. They already had an extensive basalt vessel production programme in place, from which the wheel bearings were constructed. Heretofore, basalt was used as an elite funerary material in the site of Hierakonpolis and in the various Old Kingdom examples. The potter's wheel and the Twist Reverse Twist drills were amongst the first ancient machines which used the hardest and most elite stones for new purposes in the manufacturing process rather than the end product. The Egyptians had the bureaucratic administrative means of control and redistribution of resources in order to initiate production of basalt wheel bearings should they wish to do so. By using basalt to create the wheel bearings, a prestigious stone previously restricted to the production of religious statuary to being used for industrial processes, highlights the ritual contexts and prestige for the elites who sponsored its use. The potter's wheel would therefore have been imbued with ritual prestige in its own right, and the greater skill required for learning to use it would perhaps have created a specialist potter class, albeit perhaps lower than other craft workers as suggested by textual evidence.

The royal court had long-standing trade routes with the cities of the Levant and the Near East, perhaps even some colonies in the region of Canaan (Brandl, 1992; Faltings, 1998b) and so would have had access to the pottery produced on the potter's wheel if it were traded. Strong diplomatic relations with the rulers of the city-states would have instigated the sharing of ideas as well as commodities and craft workers to teach the use of the new technology. In Chapter 5, how the potter's wheel came to be used in Egypt will be considered, and it will investigate whether the potter's wheel was used in the same way as it was in the Near East.

Chapter 5:

How did the Potter's Wheel come to Egypt?

As has been postulated in Chapter 4, if the potter's wheel was invented in the Near East, the Egyptians might subsequently have decided to borrow the idea of the invention, but then gave it a distinctly Egyptian flavour, as Baines (1983, pp. 572-599) has suggested was the case for writing. Alternatively, the potter's wheel may have been transferred to Egypt from Near Eastern centres in a form of elite technological exchange from one court to another. This proposition and the process of the potter's wheel transfer to Egypt have not as yet been addressed by scholars. In this section, the potter's wheel as an elite-sponsored technology, and whether its use commenced from elsewhere within other ancient societies will be investigated.

The invention or perhaps the realisation of the potter's wheel is considered by many scholars to be the result of a continuous evolution from a bat, mat or even calabash[29] to a turntable to a Potter's Wheel (see Chapter 4). However, the existence of bats and turntables within a pottery making tradition does not necessarily result in the eventual use of the wheel (Franken, 1971). It is possible to form a pot by rotating it with the feet while using a mat for support, a technique used by the Ibibio people of Nigeria (Nicklin, 1979). The introduction of the potter's wheel into Egypt may have been through the processes of either indigenous development or diffusion from another area, possibly the Levant. If the wheel was diffused from foreign sources, one would expect a diffusion of both the technology and the shape of the pottery (Berg, 2007). Evidence from both Egypt and the Levant in the 3rd millennium B.C. suggests that there is likely to have been contact between the two areas. There are similarities between Levantine and Egyptian pots from the 3rd millennium B.C. and this is likely since some areas of Canaan were possibly colonies (Braun, 2003, p. 24; Gophna & Van den Brink, 2002, pp. 280-281). For example, at Tell Erani (Brandl, 1989, pp. 357-388; 1992, p. 441), and Tell es-Sakan (Yekutieli, 2004, p. 171). At these sites local imitations of Egyptian pottery have been found, made of the local loess clay and showing evidence of the use of Egyptian techniques. At the same time, there is evidence that Levantine style pots were imported into Egypt. At the site of Minshat Abu Omar (Kroeper & Wildung, 1994), twenty provenanced foreign vessels have been discovered in graves dating to Naqada IIc-III (c.3650-3300 B.C.). These included wavy handled jars, ledge handled jars, looped handled jars and pots, spouted jugs, keg form vessels and loop handled jugs (Kroeper, 1989, pp. 407-419). A selected number of these foreign imported vessels have also been excavated at Ma'adi (Amer, 1936), Abydos (Petrie, 1902), Hierakonpolis (Adams & Friedman, 1992, p. 318) and Naqada (Petrie & Quibell, 1896), and are thought by the excavators to be of foreign origin. At Tell el Farkha various fragments of Palestinian bowls, wavy handled vessels, and spouted jars were found (Maczynska, 2004, p. 435). At Naqada, petrographic analysis performed by Amiran, Beit-Auch and Glass (1973, pp. 193-197) confirmed Petrie's interpretation of the vessels as imported wares (1921, p. 6). This evidence seems to corroborate the proposition that there was a long period of close interaction between Egypt and the Levant during the 3rd millennium B.C.

The Transference of the Potter's Wheel to Egypt

As has been established in Chapter 4, the potter's wheel was first utilised in the Ancient Near East c. 4000-3500 B.C., and was used to finish coil-made V-rimmed bowls rather than throwing vessels (Courty & Roux, 1995). In this section, how the potter's wheel came to be adopted or transferred to Egypt will be addressed. Any ancient society before it takes on such a new technology must be able to sustain and develop it in order for it to be a success.

There is some evidence in the corpus of pottery for the use of potter's wheel in Egypt earlier than expected, and its usage may have even occurred in Egypt at the same time as the Near East. Von der Way (1992, pp. 220-2) describes some parallels in the region of the Delta for the V-shaped bowls described in Near Eastern contexts in Chapter 4. They were found at the site of Buto in the Delta (see Map at front of text) Stratum Ia. These are described as small bowls with flat bases and thin rims with a decorative wash stripes which run parallel to the rim, and which Dessel (2009, pp. 100-1) suggests are likely to be V-shaped bowls. The decorative white wash stripes could suggest that they are "Amuq F" style smooth-faced ware, usually with "reserved spiral" decoration in red. These bowls originate from the region of Antioch in Syria, and were first discussed by Braidwood and Howe (1960, p. 232), then by von der Way who postulated that they were the result of contact between Uruk and Buto. However, examples of these V-shaped bowls occur in later levels at Buto (von der Way, 1987, pp. 247-50). Faltings (1998a, p. 23; 1998b, pp. 367-9) has suggested that there is evidence for other foreign pottery types from the Uruk such as holemouth jars, V-shaped bowls, and piecrust rims, which represented one third of the ceramic types at Buto, that are similar in nature to ceramic the corpora of the Beersheva and Ma'adi regions (Faltings & Köhler, 1996, Abb 7.1; Köhler 1998, Tafel 74.1-2; Rizkana & Seeher, 1987, pg. 47). They are made of Nile clay, which suggests that they were manufactured by Canaanite

---
[29] Also known as the bottle gourd in Jebba as part of the Nupe tradition, in western Nigeria. Nupe potters built their clay within the base of a shallow calabash gourd so that it would swivel easily (Cardew, 2002, p. 104).

FIGURE 5.1: CERAMICS FROM STRATUM IA IN BUTO. LEFT: INDIGENOUS EGYPTIAN MA'ADI STYLE HANDMADE JAR, RIGHT: URUK STYLE COIL-MADE AND POTTER'S WHEEL FINISHED V-RIM JARS MADE IN NILE CLAY. ©DAI 2012

potters living in Egypt (see Figure 5.1). Other evidence for foreign Uruk influence in the sites of Ma'adi and Buto include subterranean houses, and pottery nails for creating mosaics in mudbrick (von der Way, 1987, pp. 247-50). However, this occupation appears to be short lived, since these objects and pottery only occur during Buto Phase I (Palestinian Chalcolithic/Naqada IIa-d c.3500-3300 B.C.), and after this period Lower Egyptian style pottery and architecture predominates (Köhler, 1995, pp. 82-6).

It seems then, that at least in the Deltaic regions of Egypt, the Egyptians living in Buto and Ma'adi would have had at least visible knowledge of the pierced style of potter's wheel bearings, as undoubtedly this was the type of machine used to thin and shape the coil-made pottery. Such vessels have been coiled, then thinned and shaped on the wheel in a similar manner to V-rim bowls in the Levant. As up to a third of the pottery assemblage in Buto Stratum I contains this type of pottery, there must have been a potter's wheel in Buto at this point.[30] It is unlikely that Canaanite potters would import Nile silt to the Levant and then export back the finished pottery to Buto. However, the indigenous Egyptian population evidently did not think too highly of the potter's wheel as it did not continue to be used for pottery production after the Canaanite populace left Buto at some point around 3300 B.C. This might be due to Canaanite potters not sharing their skill with the indigenous Egyptian population. Roux (2009, p. 195) has suggested that this lack of collaboration may be due to political problems in the Levant at this time, and it would seem a reasonable premise since the Chalcolithic societies collapsed around 3500 B.C. If elite sponsorship and demand for wheel-finished V-rim bowls was removed, then possibly the potters producing them would have lost their source of employment as well as their market.

This foreign pottery influence had disappeared by Naqada IIb (3650-3300 B.C.) making way for a new local indigenous pottery with its distinctive rocker stamp design (Faltings, 1998a, pp. 30-2). It seems then, that particularly during Naqada II (c.3600 B.C.), there was a period of continuous interaction with Palestine and Sinai (Brandl, 1992, pp. 444, fig 1), even to the extent that there is evidence of colonisation of areas of Palestine e.g. Tel Erani and Taur Ikhbeineh contained Egyptian basalt and pottery vessels dating to c.3400-3300 B.C. (Oren & Yekutiel, 1992, pp. 368-373; Porat, 1992, p. 435). There is also evidence of the immigration of Canaanites into Egypt, particularly in the Delta region (Brandl, 1992, pp. 441-2). Given the interactions between Egypt and its neighbours it is more likely that the potter's wheel came from a foreign source, such as Canaan rather than Egyptian. It does appear then, that the wheel might have been invented elsewhere, probably somewhere in the Levantine regions. Egypt then firstly became an importer of the new style of pots, and then bought and transferred the technology of the potter's wheel machine together with the knowledge to create permanent workshops and kilns to support its use.

It would appear therefore that the Egyptians would have had the technological capability to create and use the potter's wheel in Egypt from approximately 3500 B.C. This is about the same time when basalt was being made into sculpture and vases (Mallory-Greenough, Greenough, & Owen, 1999), and the time when copper was being smelted at Ma'adi (Amer, 1933; 1936). However, as has been detailed above, if it was the Canaanites who brought the potter's wheel to Buto, it was not yet adopted by the indigenous Egyptians. In the next section, how the potter's wheel finally came to be used by the Egyptians will be

---

[30] One has been uncovered at Tell el Daba as a surface find, Inv no. 3379 (Bietak quoted in Arnold 1993, pg 74), however, not of the pierced type and so likely to not be contemporary to Canaanite occupation at this time.

considered and suggestions as to why this might have been will be proposed.

The development of workshop-led production

By understanding how the pottery industries developed within the Ancient Near East and Egypt it is hoped that the underlying social and economic structures can be understood. If both geographical areas had similar pottery industries based upon workshops and kilns, with wheel production run by specialist potters instigated or organised through elite sponsorship, then it is perhaps likely that the two pottery industries developed from the same model. By monopolising access to ornaments and luxury goods that require labour intensive or technologically sophisticated methods of production, and by supporting the craft specialists and production facilities necessary to create these items, elite sponsors are effectively able to control the labour market (Earle, 1987, p. 89; Shennan, 1982, p. 156). This could explain the relationship between advances in craft specialisation and the evolution of powerful elites with their associated bureaucracy to record transactions, rather than any ideas pertaining to the economic or environmental stresses which elites establish as risk managers (Peregrine, 1991, pp. 2-3; Sanders & Price, 1968; Service, 1962). Even if the Egyptians (as seems to be the case) were aware of the potter's wheel, they would have needed wealthy Egyptian sponsors to promote its use as a benefit to society, set up workshops with appropriate resources, and apprentice potters to learn the new craft.

Some scholars e.g. Childe (1954, p. 204; Foster 1959b, p. 101) thought that the idea of the potter's wheel was instigated for the mass production and standardisation of pottery. Craft specialisation is a characteristic of all known states, which can be detected in the archaeological record by examining manufacturing workshops, exchange patterns and the physical and stylistic characteristics of the goods produced by specialists (Blackman, Stein, & Vandiver, 1993, p. 61). Through constant replication and practice of the same movements it is more likely that standardisation becomes the product rather than the initial reason for developing the potter's wheel (Longacre, 1999, p. 45).

Pottery making has sometimes been viewed as a major innovation altering the course of cultural development; its invention is cited as a criterion for the transition from savages to barbarians (Morgan, 1877). Moreover, it is the potter's wheel, in particular, that is hypothesised to stimulate this technological and social transition. In Ethnographic studies, the beginning of the use of the potter's wheel coincides with a switch in pottery making between genders when the males adopt the potter's wheel while the women continue to manufacture by hand (Rice, 1991, p. 437; Vincentelli, 2003, pp. 13-4). When women do use a wheel, they tend to use it for coiling rather than throwing e.g. Danish potters of Karhuse, Island Fuenen (Vincentelli, 2003, pp. 22-3) and in Hungary during the 1880s (Szabadfalvi, 1986). In the 20th century pottery workshops and industry, taking up throwing involved crossing both a class and a gender code. Throwing was an artisan activity for men, and it was very hard to train women potters to take up the wheel and abandon hand-building, mostly due to the expense of setting up a studio. By contrast today, it is far easier to teach throwing to men and women who have no previous pottery skills (Joan Doherty pers. com. 2011). It is difficult to trace whether this might have been the problem for the Egyptians taking on the potter's wheel. Given the prevalence of women hand building potters around the world, it is easy to assume that women were the primary potters when pottery production was at the household level. Tomb scenes never seem to depict women as potters, although Wodzińska (2009a, p. 226) suspects that one of the potters shaping *dwiw* vessels in the tomb of Ty may be female. However, upon examination of the tomb, the author has noted that this figure is painted in the dark brown of males rather than the yellow colour of female bakers in the registers above[31] see Figure 5.2, but women are often depicted as bakers, brewers, hairdressers, dancers, weavers etc (Fischer, 1989, pp. 16-7; Harpur, 1987, pp. 110-4). Is this deliberate exclusion or a reflection of the reality of potter's workshop activities? Nicholson and Doherty (2014) describe the potter's workshop scenes as ethnographic depictions of potting, so we must assume therefore that the Egyptian artists were drawing what they were seeing in the potter's workshops. Perhaps these tomb scenes suggest that there was a male dominance in specialist occupations. The only examples of female potters are Nubian women making handbuilt jars using the paddle and anvil method, not attested in Egypt in dynastic times (Do. Arnold 1993, p. 21, fig 15A).

Theorists (Franken & Kalsbeek, 1975; Rice, 1987; Rye, 1981) have stated that the potter has various constraining factors to negotiate such as chemical, physical and economic before a viable pot is produced. There has been a tendency when studying ceramics to focus on the physical properties of the pot such as strength, resistance to thermal shock or abrasion, porosity and heating effectiveness (Hughes, 1981; Plog, 1980; Skibo, 1992; Rice, 1996) but social meaning was often derived from the style, shape and decoration of the pot (Berg, 2007, p. 235). This is insufficient since no account is taken of issues such as the behavioural techniques that potters use in finding clay sources or the methods they use in building their pots (e.g. through slab, coil, pinch, mould or wheel). Such issues should also be studied as they all derive from how their society is structured and organised. Traditions within the society will determine how all the underlying manufacturing processes occur, as well as what stylistic choices are fashionable at the time. Potters are adaptable to circumstances and will try to create pots from almost any clay and make the best use of whatever materials are available to them, often using a variety of everyday items as tools e.g. wire, rib bones, sponges, quills (Cardew, 2002,

---

[31] See Osiris.net for further visual confirmation http://www.osirisnet.net/mastabas/ty/e_ty_04.htm.

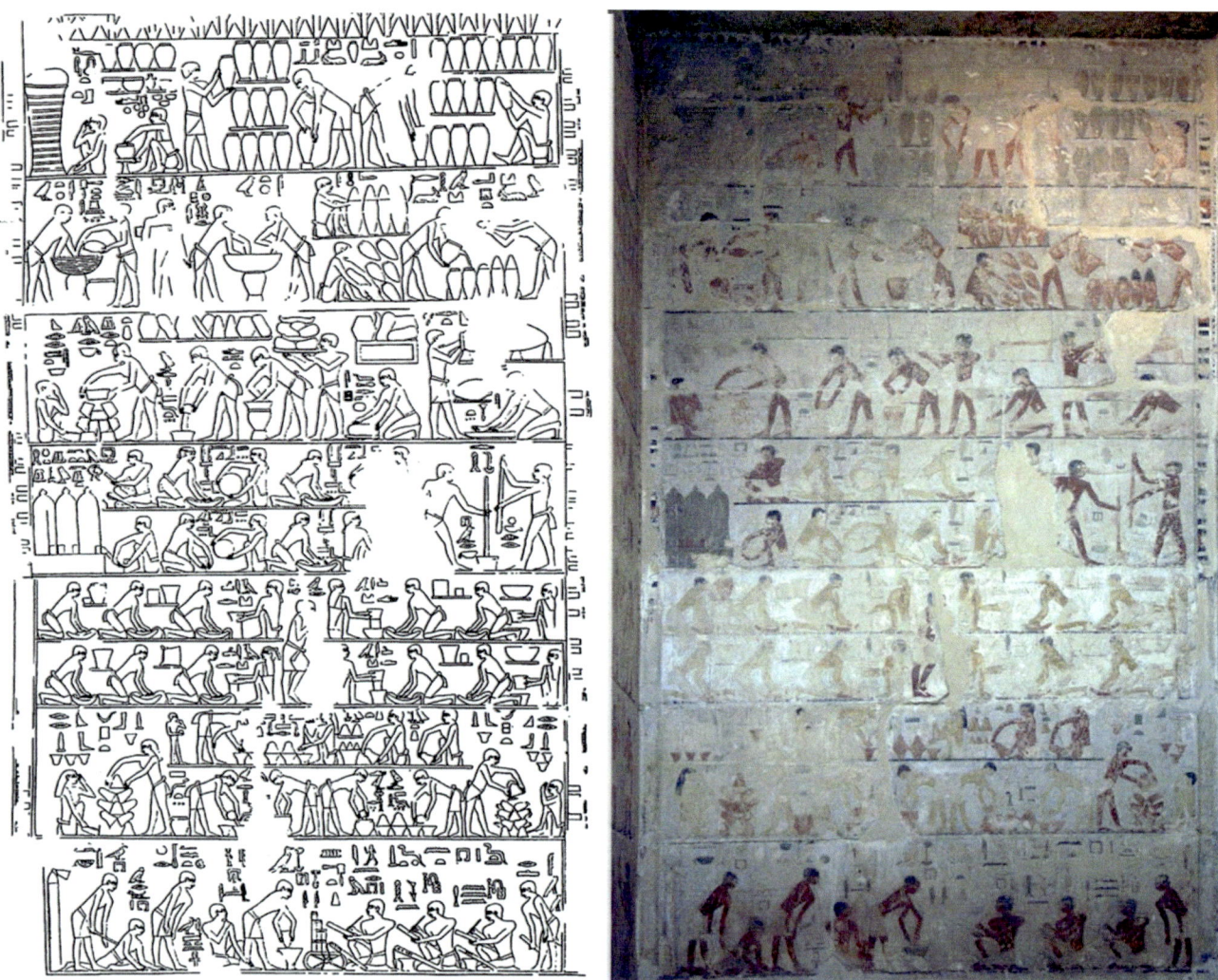

FIGURE 5.2: THE COMPLETE WALL FROM THE STOREROOM OF THE TOMB OF TY, DEPICTING THE VARIOUS CRAFT SCENES. WOMEN ARE COLOURED YELLOW, MEN BROWN. CORRESPONDENCE BETWEEN THE DRAWING FROM EPRON L., DAUMAS F.: "LE TOMBEAU DE TI", MIFAO LXV, PL. LXVI (DETAIL) AND PHOTOGRAPHS ©THIERRY BENDERITTER / WWW.OSIRISNET.NET

p. 110). In addition, they also have to subscribe to local traditions and market demands, particularly if they are producing vessels for local groups. Not having a source of a particular clay is not necessarily an inhibiting factor. For example, Cameroon potters have select areas where they gather their clays at select times of the year in gathering only one clay type, and normally their potting is restricted to during the dry season (Gosselain, 1994).

Potters who build their pottery by hand are able to use more variable types of clay as well as less refined clay bodies, whereas potters who throw on a wheel are restricted to more refined clay bodies (Colbeck, 1982, p. 10). Archaeologists often try to ascribe workshop production to the use of a particular type of clay mix or shape of vessel (Kroeper, 1992, p. 30) but occasionally can miss the points made by ethnographic researchers. For example, potters often use varied mixes of clays depending on when they have been mined or what materials they have to hand to make up a workable clay. 19th Century Balkan potters have been recorded as adding dung, goats' wool, bristles, tow, straw, chaff, soot, and/or calcite to their clay mix (Filipovic, 1951, p. 160). The suitability of the clay depends upon what the finished vessel will be used for. However ease of access to local potting clays is often the chief factor for pottery production e.g. the town of Bailén in Andelusia, Spain (Curtis, 1962, p. 491; Nicklin, 1979, pp. 441-3). In the next section, the application of such useful ethnographic studies will be applied to the advent of the potter's wheel.

Ethnographic Comparisons

As was suggested in Chapter 4, when elite individuals have control of the objects produced, they also gain control of the craftspeople and the technologies they use and therefore control the status of both themselves and the craftspeople (Herbert, 1993). Technologies allow such elite agents to construct social identities and power relations as well as producing utilitarian objects (Dobres, 1999, p. 129). The types of labour that an individual performs and the types of services or goods they offer to a group help to define the individual's place in society.

Prestige and status are often derived from the work one performs, particularly in complex societies (Costin, 1996, p. 113). Gender often establishes the range of economic activities permissible since crafting in complex societies is materially and ideologically linked to the power hierarchy and to social participation (Helms, 1993).

Ethnographic research has proved a popular method of understanding "non-western" cultures and craft traditions. Vincentelli (2003, pp. 40-4) focused her ethnographic work on a wide ranging study of the potters of the world, and postulated that the introduction of kilns and wheels nearly always involved a shift of the gender of the potters. In general, women are the primary pottery producers, making pottery by hand using the paddle and anvil technique (but see Figure 5.3), whereas the men finish the vessels or throw other vessels on the potter's wheel. When the wheel is utilised, it is associated with male full time "specialised" manufacture, whereas pottery made by hand is by women and is associated with part-time household production (Rice, 1991, p. 437).

Often the sexual division of labour is described as, "the original and most basic form of economic specialization [sic]," (Murdock & Provost, 1973, p. 203); and there is a general assumption that female craft is "domestic" whereas male craft is associated with "industry" (Maclean, 1998, p. 163). Foster (1959b, p. 113) has quoted numerous authorities on the links between men and the potter's wheel and the change in craft to more industrial processes and increased speed of production (Childe, 1954, p. 204; Harrison, 1928, p. 36; Laufer, 1917, p. 162; Turney-High, 1949, p. 174). Foster (1959b, pp. 116-7) admits that the correlation of potter's wheels and males is "one of the mysteries of history," but suggests that the differences of physical strength between the sexes may be a factor. However, as Rice (1991, pp. 438-9) contends, women seem to have no problem performing other physical labours e.g. grinding corn, laundry, ploughing etc, (see Chapter 6 for further consideration of Foster's theory). It may be that using tools or machinery such as the kiln and potter's wheel may have been taboo for Egyptian women, as has been noted by Brouwer (1987, pp. 152-153) for South Indian women as such machines were associated with the mother goddess Kali.

This gender divide occurs in most modern potteries of Egypt, as noted in the potteries of Deir el Mawas (Nicholson, 2002), and el Fustat (van der Kooij & Wendrich, 2002). However, it was not the experience of the author in 2011 at El Nazla pottery in the Fayoum where men were witnessed making hand-built and wheel thrown vessels (see http://bit.do/potterswheel and Figure 5.3). The status of potters in ethnographic studies are often viewed as lowly, as making pottery is a dirty craft (synonymous with excrement) and the pots they create are viewed as simplistic, backward containers, particularly if created by women (Lustig-Arecco, 1975, p. 6). In India, the potters belong to the lowest social class "The Untouchables" (Roux & Corbetta, 1989). In sub-Saharan Africa, blacksmiths and potters are sometimes considered to have dangerous knowledge and magical powers, and so are often kept apart from the rest of the group (Barley, 1994, pp. 63-4; Gosselain, 1999, p. 205; Herbert, 1993). This low socio-economic viewpoint of potters has parallels in the Egyptian texts described in Chapter 3, e.g. Papyrus Sallier II, column V, lines 5-6 or the *Satire of the Trades* (Holthoer, 1977, p. 18; Lichtheim, 1975). A sense of social hierarchy within the workshop can be gleaned from a variety of sources, in the form of archaeological remains, artefacts such as workshop models popular in the Middle Kingdom, literature (although the inherent bias against any occupation other than scribal has to be taken into account), and pictorial representations. Notably, in tombs and sculptures e.g. the statue from the

FIGURE 5.4: EVIDENCE FOR SOCIAL STATUS OF THE POTTER AT THE WHEEL DISPLAYING PROMINENT RIBS. LIMESTONE STATUETTE, BODY AND WHEEL IN RED/BROWN, BASE BLACK 6TH DYNASTY, TOMB OF NIKAUINPAU, GIZA, 13.2 x 6.7 x 12.5CM, ORIENTAL INSTITUTE, UNIVERSITY OF CHICAGO OIM 10628. (TEETER, 2003, P. 25)

FIGURE 5.3: A RELATIVELY RARE EXAMPLE OF A MAN MAKING POTTERY USING THE HAMMER AND ANVIL TECHNIQUE, AN ACTIVITY NORMALLY UNDERTAKEN BY WOMEN. EL NAZLA, FAIYOUM, EGYPT. PHOTO: S. DOHERTY

Priests	28	Chief stablemen	3	Incense roasters	1
Herdsmen	10	Land workers	3	Doctors	0.6
Scribes	7	Brewers	2	Guards	0.6
Fishermen	7	Potters	2	Gold workers	0.6
Coppersmiths	5	Porters	1	Measurers	0.6

TABLE 5.1: SHOWING THE PERCENTAGES OF DIFFERENT PROFESSIONS MENTIONED IN PAPYRUS BM 10068, (SHAW, IDENTITY AND OCCUPATION. HOW DID INDIVIDUALS DEFINE THEMSELVES AND THEIR WORK IN THE EGYPTIAN NEW KINGDOM?, 2004, P. 19)

tomb of Nikauinpu depicts the potter's ribs prominently displayed and clay or wooden models also suggest that the life of the average Egyptian potter was not a particularly wealthy one (see Chapter 3 and Figure 5.4).

Textual sources also provide useful information relevant to the life of the potter. Shaw (2004, pp. 18-19) has listed the professions of 182 households of "the town region of the west of No from the temple of Menmaare to the settlement of Maiunehes," found on the verso side of Papyrus BM 10068, one of the collection of *Tomb Robbery Papyri* dating to the year 16 and 17 of Ramesses IX (see Table 5.1). While the papyrus is of New Kingdom date, it is still a valuable resource. The high number of priests is likely to be due to the proximity of a temple in the area, probably the temple of king Menmaare that is mentioned in the text. It is also likely to be indicative of the situation on the West Bank of Thebes, which this text describes. The relatively low number of the potters in this case would make sense when compared with modern ethnographic research (Nicholson, 1995b). Often a potter would have several assistants, frequently children, and only a few of these would have considered potting to be a full time activity, since this was restricted to certain times of the year when clay sources were good or when the fields did not require tending e.g. inundation.

In this case, it is likely that the potters identified are those who specialise in using the wheel rather than the potters making pots by hand, since hand-made pottery would have been done as and when it was required by the household. As is noted ethnographically, women could also have produced household pots, perhaps undertaking this activity alongside other chores. While women almost are never observed working on the potter's wheel in traditional workshops, they will perform almost all other tasks including clay preparation, hand building, decorating, and finishing etc (Nicholson 1995b, pp. 283, fig 9.2). In addition to the knowledge and skills, the use of the wheel required rather more organisation and continuous use for it to be a profitable venture and to enable ongoing supply to a demanding market. Models and tomb scenes indicate that there may have been particular areas where workshops were located, often close to other craftworkers' workshops in industrial quarters e.g. carpenters or blacksmiths (see Chapter 3 Table 3.1.) as seen at the Amarna (Nicholson 1992; 1995a) and Lachish excavations or near to temples and palaces as at Hierakonpolis and Tell Yarmuth (Baba, 2006; Roux & de Miroschedji, 2009; Roux, 2009, p. 199). Whether all potter's workshops were near to palaces and as part of the estate of wealthy landowners is uncertain, but some archaeological remains indicate that some craft activities, notably potting, cobbling, painting, and bread making could all have been performed at some level in the home. The sheer quantity of basalt chippings found at the Amarna houses (P49. 3-6) would provide evidence of a series of workshops next to the houses (Kemp, 1995).

Longacre's (1999, p. 44) notion that the more specialised an industry and a person becomes, the less creativity they display can be shown to be true from ethnographic studies, as the pottery making is reduced to a production line not dissimilar to our mechanised systems. The modern potters of Ballas have organised the production of their amphorae into clearly defined activities. The Ballas potters make their wares using marl clay from one part of Egypt, Ballas (Deir el Gharbi) in the Delta (Nicholson & Patterson, 1989). Compared to the Nile-silt potters' chaotic and disorganised workshops, the Ballas industries appear highly organised with mostly all of these potters making amphorae in large workshops (Bourriau, 2002, pp. 78-95). There are often several workshops in one area, all being supplied with the same clay by miners. It is possible to trace which potter made what type of pot, as each potter's apprentices form the handles of the amphorae in a particular way. In addition, they allocate particular intermediaries to sell their wares to a specific market, each market favouring slight differences in the size, shape, texture, or temper of the pot. This is the result more of taste and fashion rather than particular obvious differences (Nicholson & Patterson, 1989; Nicholson, 2002, pp. 138-146). Similar inferences could also be made for ancient potteries, where presumably the ancient market was just as fickle with its own preferences. Ancient sources provide evidence of the chaotic nature of some potteries and the relatively low status of some potters, if they were not a master craftsman.

Kilns, Potter's Wheels and Workshops

When using a potter's wheel, potters need to be located in a permanent workshop near clay sources and local markets (van der Leeuw, 1976, p. 87). For practical reasons, the potter's wheel given its weight, and size (see Chapter 2), needs to be embedded into the ground in order to be spun effectively and avoid oscillation. The use of the potter's wheel would mean that the entire pottery production process must become more mechanised, with apprentice potters assisting the lead potter by handing him clay as he works to speed up the production of pots. This rate of production would allow for the extra cost of transporting finished vessels and procuring clay through dependable transport.

Hand-building and open firing have arguably no need for structures or permanent features such as kilns or workshops, but require only a suitable source of clay and firing material, and consequently archaeological visibility would be very low or non-existent. In contrast, one would expect kilns and areas of workshop activity to be more obvious to archaeologists as the continual firing of the kilns would show up in magnetometry surveys e.g. at Gurob (Boatright & Hodgkinson, 2010; Hodgkinson, 2012) and workshop floors and walls should also be more easily discernible. There are relatively few early pottery workshop sites identified in Egypt, with the exceptions being at Hierakonpolis (Adams, 1974, p. 20), and Abu Sir (although a variety of kilns have been uncovered see Appendix I). In the section below, the evidence for the development of pottery kilns, workshops and wheels will be investigated.

The organisation of pottery workshops has been variously discussed elsewhere (e.g. Costin 1991; Peacock 1981; Rice 1987:183-91; van der Leeuw 1976 and in Chapter 4). The different steps and/or complexity of the production process suggest different levels of ceramic production. Consumption of products such as ceramics is assumed to be at the local level, due to limitations of transport, market demands, and ease of production. In most ancient societies, the basic level of organisation of production was focused on the household, for reasons of self-sufficiency (Sahlins, 1972). Van der Leeuw (1976) defines pottery household production as occasional, simple and produced locally by non-specialists, and Rice (1987, p. 184) adds that the household system has little opportunity for specialisation and intensification. For household production, the most technologically simple methods were employed with little investment in specialist machines or tools. Often clay fabric types in use locally relate to function e.g. limestone chips were added to the clay used to create to cooking vessels and such vessels have been found in the Southern Levantine sites dating to Early Bronze Age I (Dessel, 2009, p. 122).

The next step up is household industry where the production of pottery is made by semi-specialists where the finished products had an exchange value and were produced solely for this purpose. An example of this could be wavy-handled jars, black-topped and red-polished, made in the Pre and Early Dynastic Periods specifically for depositing in graves, which were often filled with mud or ash rather than food (Serpico, 2004, pp. 1018-9). Kiln sites near the Great Wadi at Hierakonpolis seem to suggest specialist production outside of the domestic sphere (Friedman, 1994, p. 896) but perhaps on an part-time level. Given that the products were mostly consumed by the communities who made them there is little evidence for long distance exchange as during the Naqada I and II (Takamiya, 2004, pp. 1034).

Full time specialisation is much more easily discernable in the lithics and stone vessel industries than in the ceramics. By the Naqada II period (3600-3200 B.C.) there was a rapid expansion in the volume of stone vessels being produced, perhaps suggesting that a faster and more reliable method of production was introduced. This was particularly the case at Hierakonpolis, where stone vase workshops have been located less than 50m away from the main temple of Horus of Nekhen (Kemp, 2006, pp. 196, fig 68; Quibell & Green, 1902, p. plate LXXIII see Chapter 4, Figure 4.10). Another workshop dating to the Old Kingdom has recently been located near Sheikh Said (Willems, et al., 2009) perhaps indicating that the temple personnel were regulating the vessel production. Secondary evidence demonstrating the control of elites by organising the craft workers can be found in the increased use of administrative paraphernalia such as writing systems, labels, potmarks (Wodzińska, 2009b) and seals (Baines, 2007). Long distance trade networks involving these craft specialists were likely to be actively controlled by the rulers and their court, and is represented on objects such as the Gebel el Arak knife[32] (Mark, 1997). However, the term specialisation for mass-production or commercial purposes cannot be applied to pottery during the Naqada III periods, since at this point the specialists were importing and working luxury goods e.g. ripple flaked knives, Decorated ware pottery (Aksamit, 1992, pp. 17-21). The value of such items was derived from the elaborate nature of their manufacture by craftworkers attached to the temples. The elites no doubt organised this so that they could control what the artisans were making, but they may also have wanted to create social inequality (Brumfiel & Earle, 1987, p. 3). The introduction of the potter's wheel would be the next logical step in this increasing specialisation process. However, before it could be fully incorporated into the workshop and developed for more industrial means, it would first have to be accepted as a beneficial technological improvement which would enable the elites to control the potters working in their estate and the temple workshops and therefore merit elite sponsorship.

The potter's wheel and the kiln seems to make their earliest appearance pictorally in the 5th dynasty mastaba tomb of Ty at Saqqara (see Chapter 3, Figure 3.2; Steindorff, 1913, p. tafel 83 and 84). As discussed previously, there is much evidence to suggest that kilns, potter's wheels and indeed workshops were likely to have been in place from Predynastic and Early Dynastic times. In the Naqada III period, pottery made in Upper Egypt was being distributed to Lower Nubia and southern Palestine, indicating the establishment of pottery workshops which produced vessels to hold the products being traded (Brandl, 1992). Until the Naqada II period, pottery was exclusively made

---

[32] Musée du Louvre, (Accession number E 11517).

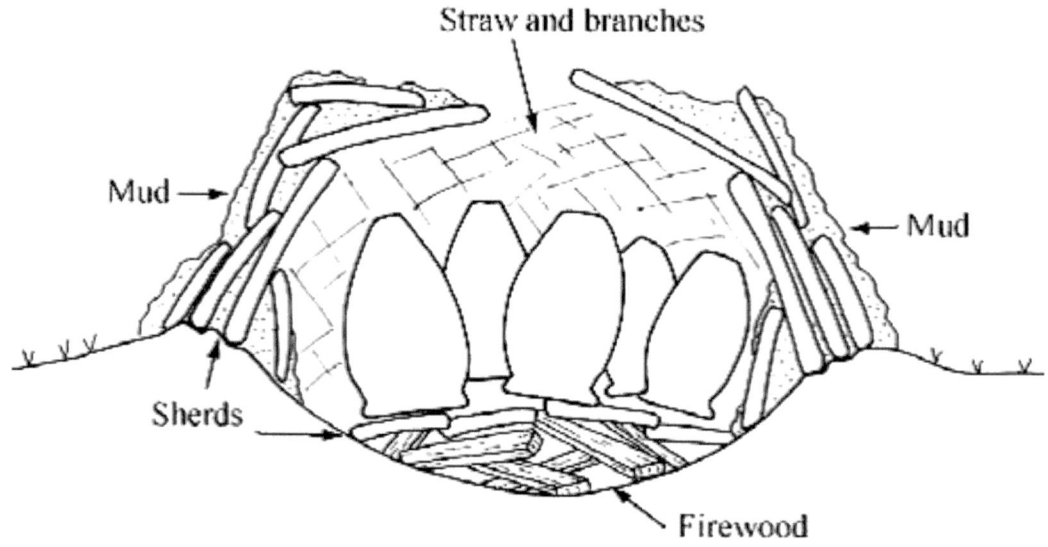

FIGURE 5.5: EXPERIMENTAL RECONSTRUCTION OF THE PIT KILNS LOCATED AT HK11 C SQUARE B4NW (BABA, 2005, P. 20)

of alluvial clays.[33] These clays are easy to work, shape and fire and are the most ubiquitous clay in Egypt as they can be relatively easily collected from all along the alluvial plain of the Nile. The beginnings of the use of Marl clays, mostly only available from the Ballas and Qena regions of Egypt, and their generally higher firing temperature perhaps meant that the Egyptians had to become more organised in their pottery production. During the Old Kingdom on the tomb of the deceased person it was popular to describe common everyday life scenes that the deceased might have been associated with during their lifetime. At this time, the elite members of the society became more of a person in their own right rather than just an extension of the Pharaoh's court as had previously been the case when the courtiers were buried beside their king in secondary burials. Ty was Director of the Hairdressers of the Great House (i.e. the palace) and also the overseer of estates and temples of Kings Sahure and Neferirkare and so would have been involved with the day-to-day administration of the temples and estates. Presumably, his duties included organising the supply of pottery, and its production, although indirectly. The display of the workshop scene is therefore quite pertinent. Ty's workshop seems to provide evidence that there might have been specialised potters who were involved in the making of select pottery shapes.

It has been noted by Wodzińska (2009a, pp. 233-239; 2009b, p. 245) that conical and flat bread moulds have different potmarks at the 4th dynasty site of Giza, even though all were made at the same place. These could be linked to different workshops, or being made onsite in bakeries. In modern pottery production potters specialize in particular shapes and often produce only a set number of vessel shapes, despite being capable of more (Nicholson,

2002; Wodzińska, 2009a, p. 237). Nile silt clay potters in contrast to marl clay potters seem to produce a more varied corpus (Nicholson & Patterson, 1989; Nicholson, 1995a, p. 294).

The firing of pottery in Ancient Egypt occurred in a variety of ways. Arguably the simplest was bonfire firing, clamp kiln or open firing where a trench was dug, fuel and sawdust laid inside and the pots arranged in it with more fuel placed on top. Once lit, there is relatively little control over the temperature of the fire, but provided that pots are well arranged and not too much fuel is used, many viable pots can be produced. The next kiln stage is the more sophisticated screen, pit or box kiln, known from Predynastic times and found at site HK11 C at Hierakonpolis, which may have been used as an aid in the brewing process rather than a kiln (Friedman, 2004, p. 18; Baba, 2006, p. 19; Takamiya & Baba, 2004, p. 19). At Hierakonpolis, these screen kilns had walls of mud brick c20-30cm surviving in a square or horse-shoe shape surrounding a mud brick platform and 2-2.5cm in diameter. Pots were placed in the back of the pit with the fire at the front to take advantage of the prevailing northerly winds (see Figure 5.5). Some examples of these kilns are partially buried in a pit and then the walls built around this, most pits are 1m in diameter (Baba, 2005, pp. 20-1).

Some "large pot" kilns such as that found at el-Mahasna (Garstang, 1902, pp. 38-40) and square A6 at Hierakonpolis contained rods of clay apparently being used as fire bars or fire dogs c. 15-25cm high. These were placed in an upright position, either to support the pots and keep them away from the flames (Takamiya & Baba, 2004, p. 19) or to support larger vessels in which smaller ones were placed to protect them from the excessive heat (Harlan, 1982; Nicholson, 1993, p. 108). Feature 12 at HK11 C (see Figure 5.6) contained 13 such firedogs, standing upright, slightly

---

[33] Which Egyptologists refer to as Nile Silt, see Appendix II.

FIGURE 5.6: THE FIRE DOG FEATURES FROM HIERAKONPOLIS SQUARE A6, HK11 C, FEATURE 12. (TAKAMIYA & BABA, 2004, P. 19)

curving inwards arranged in four concentric circles, which became successively smaller. Each circle of firedogs were of different heights to support a vessel. These sort of "large pot" kilns are likely to have been in use from Naqada II onwards c.3650 B.C. (see Figure 5.7).

The famous Black Topped ware (see Chapter 4, fig 4.2) made of thin walled Nile silt clay may have been produced in screen/box kilns similar to this (see Figure 5.6 and Figure 5.7), indicating the beginnings of workshop production and specialisation of potters. In order to achieve the black and red colour differences, it seems that the Predynastic potters created a combination of reduction firing and carbon rich smudging, perhaps initially invented to make the interior of the vessel less porous to water than the outer. This was likely to be a one step, rather than a two step process as suggested by Lucas (1932, p. 94) with the pot being upturned and sawdust or resin placed in the base to induce carbon smudging (Hendrickx, Friedman, & Loyens, 2000, p. 173). So far no kiln has been identified particularly with black topped vessel production (Hendrickx, Friedman, & Loyens, 2000, p. 178) despite Barbara Adams (2000, p. 20) suggesting the kiln sites at cemetery 6 as likely candidates. It is possible that the majority of Predynastic pots were fired in a bonfire style firing (Spencer, 1997, p. 46). Although the area did contain a number of Black topped pottery wasters (Smythe, 2005, p. 21). Maczynska (2004, p. 428) considers that the use of screen kilns are likely for Petrie's P-ware (polished red) and s-ware (late class) as they are of finer fabric. In addition, kilns Hk39, Hk40, Hk59 and Hk59A at Hierakonpolis have been excavated next to red polished ware jars (Hendrickx, Friedman, & Loyens, 2000, p. 176; Geller, 1984, pp. 92-94).

From the Old Kingdom onwards the most common kiln was the updraught kiln, first depicted in the pottery workshop scenes of Ty in the 5th dynasty (Chapter 3, Figure 3.2 and Figure 5.2 this chapter). Box kilns were still in use up to the New Kingdom and most likely beyond and some examples have been identified in Amarna's industrial quarter by Nicholson (1989). Updraught kilns are tall, biconical, circular or horseshoe shaped kilns ranging from c0.8m-3m in diameter (see Appendix I). Their larger size and more controlled firing capabilities may have allowed the Egyptians to experiment with their pottery wares and to use desert Marl clay that generally seemed to be fired at slightly higher temperatures than Silts. In addition, the new style of kilns may have allowed the Egyptians to use finer pastes of clay for their pottery vessels, while increasing the likelihood of more finished vessels surviving the firing process (Nicholson & Patterson, 1989; Nichsolson, 1993, pp. 105-106). These larger kilns ensured greater fuel efficiency with less heat loss through the walls, a higher temperature and better control of the atmosphere around the pots (Hodges, 1971, pp. 35-39; Wood, 1990, p. 26).

Updraught kilns are a much more permanent structure than bonfire firing since they comprise a circular structure of walls of mud bricks with a firebox in the centre, separated by a partition with a perforated floor supported by a central

FIGURE 5.7: THE SCREEN KILN AT EL MAHASNA (GARSTANG, 1902, P. 39)

FIGURE 5.8: THE ASSISTANT POTTER IN THE TOMB OF TY IN FRONT OF THE KILN, SHIELDING HIS FACE WITH HIS HAND. STOREROOM, REGISTER 7, SAQQARA, EGYPT C.2450-2300 B.C. (ÉPRON & DAUMAS, 1939, P. PL 71)

wall or pillar. A fire is constructed in front of the firebox, hot gases rise and pass through the perforated floor and into the vessels and then out through an upper chimney or flue. Temperatures are controlled by the intensity of the fire and the amount of air draft allowed in. Sometimes a screen is affixed to cover the air hole (see Potter 1 in Chapter 3 Figure 3.2 and Figure 5.8 above) where the potter has opened up the kiln door and is screening his face from the heat). Early examples of updraught kilns have been uncovered at Ain Asil, where various kilns and two associated workshop remains were uncovered southwest of the main town (see Figure 5.9). The kilns belong to four phases of use, with most fireboxes opening to the south, perhaps to take advantage of the prevailing winds (Soukiassian, Wuttmann, Pantalacci, Ballet, & Picon, 1990, pp. 5-9).

The fact that such a large structure as a kiln was needed suggests that pottery production from the Early Dynastic Period became a more industrialised process, with permanent workshops and specialised workers i.e. the potters required to work all day every day solely to produce pots. There was clearly a demand that needed to be met beyond domestic household requirements. The use of the potter's wheel may have been fundamental in this process. It seems to be no accident that the first depiction of the potter's wheel in Egypt in the tomb of Ty (Épron & Daumas, 1939) also includes the first depiction of a pottery kiln, and a pottery workshop, and consequently provides clear evidence of elite sponsorship. Within this workshop at least four specialist potters and their assistants are all engaged in work to produce pottery quickly and efficiently.

When then, does elite sponsored-workshop production become more widespread in Egypt and how does the potter's wheel fit into this process? Archaeologists often ascribe higher value to elite "luxury" goods e.g. fine

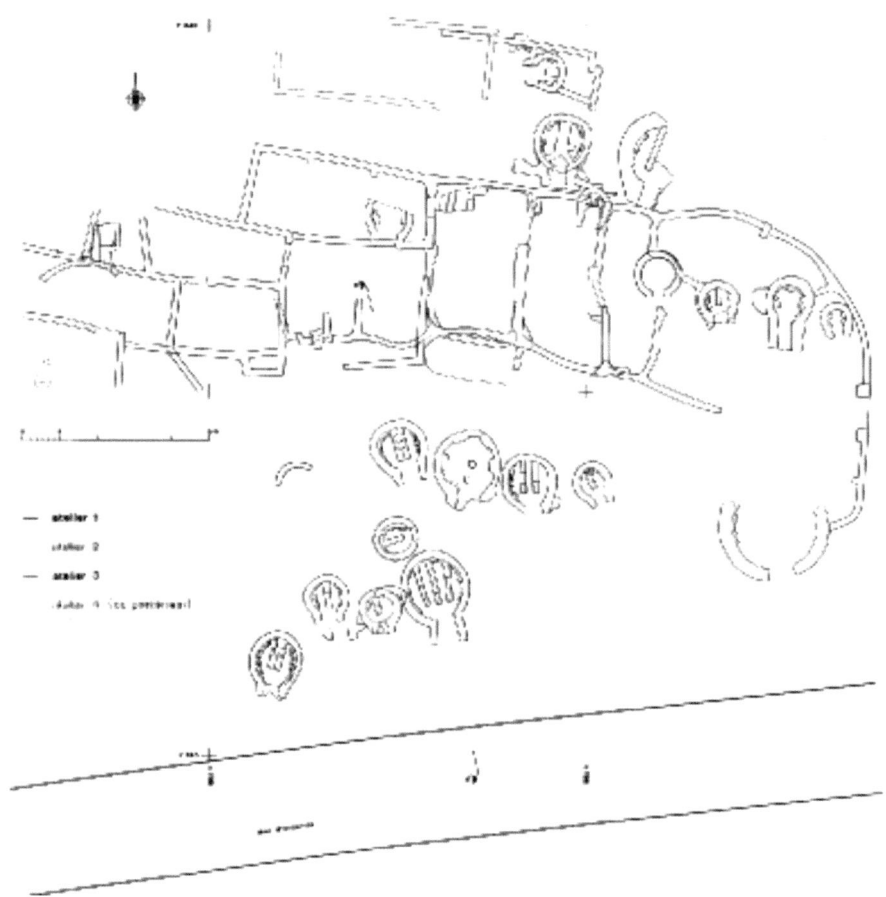

FIGURE 5.9: THE MULTI-PERIOD POTTERY WORKSHOP AT AIN ASIL. SOUKIASSIAN ET AL. (1990), PG 12, FIG 5

tableware pottery. However, in antiquity, greater intrinsic value may have been ascribed to the more utilitarian items such as storage jars used to transport staples such as oil, grain etc (Bourriau, 2002). So called elite items were usually destined purely for the grave, and were likely to have been made by local artisans, whereas storage vessels e.g. amphorae would have been transported throughout the known world and would have been made on an industrial scale in workshops using potter's wheels designed for the task.

Workshop specialisation

Although there is no single definition of craft specialisation (Rice, 1987, p. 281), it is often viewed as the standardisation of vessel shape, size and fabric. However, such studies often focus on the results of production, namely the pots rather than the means of production, i.e. the workshop and tools used to create the vessels (Dessel, 2009, p. 124). Some of the best examples of pottery workshops come from the Near East rather than Egypt (see Chapter 2, with the possible exceptions of Abusir (Verner, 1992), Amarna (Nicholson & Patterson, 1989; Nicholson, 1995b: 1992) and Ayn Asil (Hope, 1979; 1995; Soukiassian, Wuttmann, Pantalacci, Ballet, & Picon, 1990)). These examples enable insight into the industrial processes and production methods employed by the Egyptian potters, with tomb scenes such as those at Beni Hasan and written records e.g. Neferirkare and Raneferef archive (Posener-Kriéger, Verner, & Vymazalová, 2006, pp. 266-268, pl. 48-9), Abu Sir (see Chapter 3) being used to fill in the gaps (Nicholson & Doherty, forthcoming).

At the Late Bronze Age IIA workshops at Hazor excavated, by Yadin 1958 and 1960, a series of buildings associated with a cultic shrine contained a potter's workshop with a workbench, potter's wheel bearings, cobbled floor clay preparation areas and tools. Three sets of wheel bearings were all found within a larger potter's quarter. The workshops had open-fronted booths on the streets perhaps for the potters to sell their vessels. One of the wheel bearings was uncovered in association with a cultic pottery mask, perhaps indicating the cultic significance of the pottery produced in the workshops (see Figure 5.10). The workshops were located close to a stelae shrine and the Hazor city ramparts.

The Canaanite potter's workshop dating to the Late Bronze Age III located in a cave near the tell at Lachish (Tell el-Duweir Palestine) was discovered in the 1937/8 season by J. L. Starkey and contained one of the most complete set of potter's materials and instruments (Torczyner, 1938; Tufnell, Murray, & Diringer, 1953). Notable among the finds were two in situ potter's wheels, comprising two stones of basalt and local limestone. This workshop and its contents helped to answer a great many questions regarding the methods and tools that the potter used. The workshop was located within a cave relatively far from the city residential areas, and close to good sources for the procurement of clay (good sources of loess clay were

FIGURE 5.10: HAZOR POTTERY MASK (C 1136) AND WHEEL BEARING (C1200/2) IN SITU. LOCUS 6225, STRATUM IB LB II YADIN 1958, PL CLXXXII AREA C

located approximately 180 metres away (Rosen, 1986, pp. 58, 129-31)) and water. In addition, the resulting smoke of the kiln would have made it undesirable to other residents. In the hot climate, the cave would ensure that the clay remained cool, and the isolated location would enable the potters to work in peace without much distraction and they also would have a wide area in which to dry their pots. It is interesting that one of the potter's wheels (or mortar/pivots as described in the site plan see Figure 5.11) are located apart from the rest of the workshop, in pit A. At Lachish, the Late Bronze Age potters' tools included bone points, pebble and shell polishers, sherd smoothed to use as ribs or turning tools were uncovered (Magrill & Middleton, 1997, pp. 68-9,72, fig 6a; Tuffnell, 1958, pp. 291-3, pl 49:12-13).

These two examples of pottery workshops suggest that the production process became more industrialised during the Late Bronze Age IIA and III periods. They were located on the outskirts of towns as in the Lachish example, possibly due to the low status of the potters, the associated fire risk of the kiln and the easier and quicker access to clay and fuel that the potters would have by not being inside the town (Simpson S. J., 1997a, p. 50). Within the site of Lachish, there is perhaps evidence of craft specialist

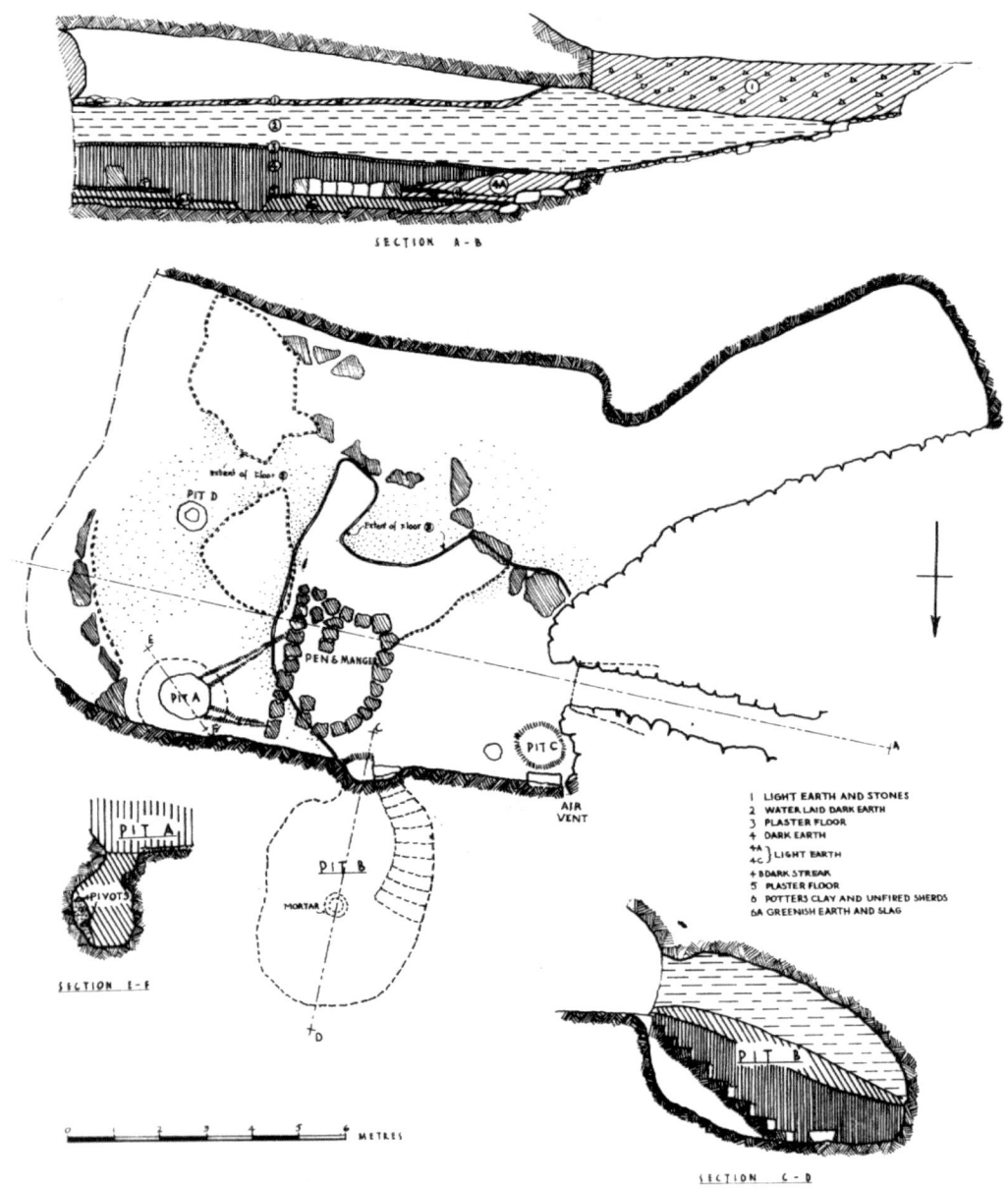

FIGURE 5.11: PLAN AND SECTION OF CAVE 4034 AT LACHISH (MAGRILL & MIDDLETON, 1997, P. 69)

segregation, with the master potter having his own designated area and the apprentices occupying the rest of the workshop space. Pit B (located down a flight of steps) was where the results of the potter's labour were left to dry and around forty complete fired vessels were stored, close to both the entrance and the wheel, presumably so that the master potter could oversee the working environment and processes. The potters were apparently supplying their local community located in the town upon the tell at Lachish. Some may also have been produced for the wider trading markets (Magrill & Middleton, 1997).

The First Occurrence of Wheel thrown Pottery in Egypt

It seems that it is no accident that the first examples of wheel thrown pottery comes from contexts of the highest social level: that of Egyptian royalty. Beginning with the reign of Pharaoh Sneferu, first ruler of the fourth dynasty (c.2640-2604 B.C. father of Khufu who built the Great Pyramid at Giza), small arguably inexpertly thrown vessels began to appear in royal cultic contexts. They were mass-produced, but were destined for a specific cultic and funerary sphere, and were used daily as part of the offering rituals to the dead and then discarded after one use (in a similar manner to the Mesopotamian bevelled rim bread bowls (Beale, 1978)). These 4th dynasty miniature vessels seem to only occur in elite and royal ritual contexts, such as pyramid mortuary temples, tombs and chapels, beginning with the foundation deposit at the pyramid of Sneferu at Meidum (Clayton, 1994, p. 45; Dodson, 1995, p. 27). These vessels were created to serve and nourish the ka of deceased royal and private individuals with a token offering of food and drink, shaped to look like miniature plates, beer jars and uguent pots (see Figure 5.13). Once

How did the Potter's Wheel come to Egypt?

FIGURE 5.12: THE MINIATURE VESSEL DUMP OUTSIDE SNEFERU'S MEIDUM PYRAMID, 4TH DYNASTY FROM EL-KHOULI'S 1991 EXCAVATIONS. PHOTOS: S. DOHERTY

FIGURE 5.13: EXAMPLES OF MINIATURE VESSELS FROM MEIDUM © PETRIE MUSEUM OF EGYPTIAN ARCHAEOLOGY (PETRIE 1892, PL XXX; PETRIE, MACKAY, & WAINWRIGHT, 1910, PL XXV). UC17625 9.3CM, UC17632 7.4CM, UC 17630 6.7CM

the rituals were finished for the day, the used vessels were then deposited into large rubbish pits such as those at the pyramid temple of Menkaure found by Reisner (1931). It is likely that later they ended up being incorporated into wall linings e.g. mastaba of Prince Neferma'at at Meidum or foundation deposits when the pits were cleared to make way for more (Charvát, 1981; Reisner 1934) (see Figure 5.12 above).

Excavations at the edge of the Meidum pyramid of Sneferu in 1989-1991 (el-Khouli, 1991) revealed vast quantities of miniature vessels within the debris of previous

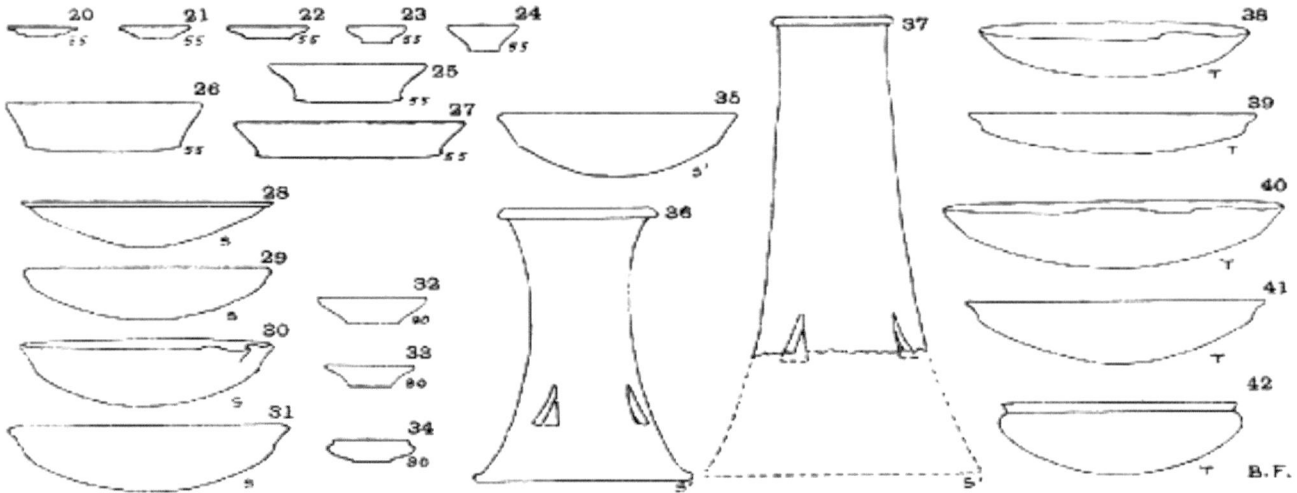

FIGURE 5.14: THE MEIDUM PYRAMID FOUNDATION DEPOSIT, CONTAINING MINIATURE VESSELS (NOS 20-24), EXAMPLES OF MEIDUM BOWLS (E.G. NO. 42) AND BASALT STANDS (36 & 37) WITHIN A SEALED 4TH DYNASTY CONTEXT. PETRIE, MACKAY, & WAINWRIGHT, 1910, P. 2 PL XXV

excavations (Flinders Petrie 1890-91, published 1910), Pennsylvania University under director Alan Rowe in 1929-30 (published 1930)) close to ground level, which former excavators had left. el-Khouli's (1991) team uncovered part of the foundation walls of the approach of the small mortuary temple previously exposed by Petrie's excavators (1910, p. 2). Alongside the walls hundreds of miniatures were discovered, mostly complete (el-Khouli, 1991, p. 13). More were found during excavations at the nearby mastaba of Neferma'at and Itet, together with some rim and body sherds of Meidum bowls, all dating to early 4th dynasty (Milward Jones, 1991, pp. 43-5).

Petrie was the first to uncover the mortuary pyramid temple of Sneferu, the causeway leading up to the temple and an unfinished causeway known as "the Approach," which runs by the edge of the southern side of the causeway stone foundations and contains the remains of a retaining wall. By this wall, two foundation deposits of 21 items of pottery, pottery and basalt stands, stone vessels and a corn grinder were uncovered dating to the late 3rd and the early 4th dynasty (Petrie, Mackay, & Wainwright, 1910, p. 2 pl XXV). The foundation vessels were comprised of miniature pottery bowls, stone model vessels or Meidum vessels, some spouted. No other pottery types were found within the foundation deposit, highlighting the very great significance that these royal funerary vessels had within the ritual (see Figure 5.14). Significantly, it seems that Sneferu was the first Pharaoh to instigate the sponsorship of potters to use the potter's wheel for throwing miniature vessels destined for the cult of his *ka*. Perhaps these potters even lived within the vicinity of the Meidum pyramid town *t't šneferu* (see Chapter 7).

Summary

In this Chapter, the investigation of the conditions necessary for the uptake of a new technology has enabled greater understanding of why the Egyptians adopted the use of the potter's wheel. There is evidence of close interaction and sharing of pottery types and new technologies throughout the Near East, Levant and Egypt. The potter's wheel it appears was first used in the Near East between 4000-4500 B.C. There is evidence for the potter's wheel in the Egyptian delta c.3500-3300 B.C. as Canaanite potters living in Egypt used a potter's wheel to thin and shape Canaanite style pottery but using local Egyptian Nile silt clays. As in the Near East, the instigation of the potter's wheel was through elite sponsorship, possibly through transference of potters between the royal courts, but more likely through colonisation of Canaan and Palestine; and through trade links between Egypt and these neighbours. An analysis of where wheel thrown pottery occurs (whether that be in domestic, funerary or cultic contexts) and where pottery workshops have been located has aided recognition of the development of the kiln, the potter's wheel and the pottery workshop as a potentially elite-sponsored craft undertaken for a specific purpose other than the mass-production of domestic wares. The potter's wheel was used during the Middle Kingdom Egypt to manufacture mass-produced wheel made pottery. However, initially the potter's wheel was initially used to produce a select range of miniature and model vessels within particular context, of funerary and cultic offerings (Bárta, 1995, pp. 22-4). Several thousand of these miniature vessels have been found particularly in contexts such as the pyramid temples of Sneferu at Meidum (Allen, 2006, pp. 19-21), Menkaure at Giza (Reisner, 1931, p. 228) and mastabas such as Ptahshepses at Abu Sir (Charvát, 1981, p. 148).

There are gender issues relating to the use of the potter's wheel, supported by ethnographic parallels. When males start to manufacture pottery using the potter's wheel, it is a full time specialised activity designed to manufacture pottery destined for prestigious elite contexts. Women do not appear to have utilised the potter's wheel, but seem to restrict their potting activity to part-time hand-building vessels for household use. Men seem to work in pottery workshops, either located in industrial quarters with other craftworkers or near to temples, palaces or shrines. The levels of ceramic production were detailed including the development firing processes involved, and it appears that by the 5th dynasty the most common form of kiln was the updraught kiln. This is the most likely choice to be used when potter's wheels were incorporated into the pottery production sequence, as the two are always depicted together in tomb scenes.

Chapter 6 will address how the Egyptians went about this new process of throwing vessels on a potter's wheel. Heretofore, the potter's wheel had only been used by the peoples of the Near East (and also in the colonies around Buto in the Delta) to finish and thin coil made V-rimmed pots (Faltings, 1998b; 1998b). Throwing however, is an entirely different process, requiring the potter to learn different bodily movements and new skills in order to achieve the desired result. Chapter 6 aims to deconstruct the manufacturing methods used by the Egyptians to create wheel thrown pottery.

Chapter 6:

Detecting the Use of the Potter's Wheel in Egyptian Pottery

By examining manufacturing marks on pottery and determining which marks are characteristic of wheel-made wares through comparing them to experimental examples, it is hoped to achieve a more complete view of when and in what manner the Egyptians were manufacturing their pottery vessels on the potter's wheel. The first step (as begun in Chapter 5) will be to identify possible wheel thrown pottery through examination of Egyptian pottery collections of the Petrie Museum of Egyptian Archaeology, the British Museum, the Ashmolean and Cyfarthfa Castle to consider to what extent the use of the potter's wheel can be noted on pottery. Some of the Predynastic and Early Dynastic pots from these assemblages seemed to display certain characteristics associated with wheel thrown pottery, whereas pottery of the Old Kingdom exhibited still more. A set of criterion will be outlined in this chapter based on experimental and archaeological examples of wheel thrown pottery.

Through practical experimentation by manufacturing replica pottery using a reconstructed potter's wheel based on pictorial, literary, ethnographic work and excavated potter's wheel bearings (as outlined in Chapters 2 and 3) it will be possible to deconstruct the manufacturing methods used by the Egyptians to create wheel thrown pottery by comparing them to modern throwing techniques. From these experiments, a greater understanding of how to determine what manufacturing processes were involved in the excavated pottery assemblages will be achieved. This chapter aims to provide a fresh perspective to analysing and examining wheelthrown pottery and to gain a greater understanding of the techniques that the ancient potters used when making their pottery using the hand-spun potter's wheel.

The journey towards the invention of the potter's wheel is likely to have evolved in a number of sequential stages. A supporting device e.g. a large pot sherd or mat rotated on the ground may have evolved into a turntable using two pierced basalt disks, and then into a potter's wheel. Alternatively, a potter may have decided to develop their own potter's wheel, using the already known pivot and socket designs which were formerly used for door hinges and known as early as Naqada II (see Chapter 4, Figure 4.9). Whatever the method used to obtain or invent the wheel, the crucial development was the expertise and capability of the potter to utilise centrifugal force in order to achieve a pot that was thrown. Ethnographic evidence such as that relating to the potter's of South America (Litto, 1976, pp. 106-7), suggests that it is possible to rotate a pot on something as simple as a large pottery sherd, so that it exhibits rilling marks similar to wheel-made pots. These spiral and rilling marks usually only occur on the uppermost part of the vessel when the base of the vessel is sufficiently dry to rotate it at speed. The base is placed in a depression in the ground and rotated allowing the body of the pot to be shaped through centrifugal force, but without the centring effect of a potter's wheel (Shepard, 1968, pp. 60-62). This methodology is exemplified in early Naqada IIb pots which often show fine parallel rilling marks around the inside of their rims and upper bodies, but rarely on the bases or lower parts of the vessels. The wavy handled jars, in particular, appear upon consultation to have been built up in coils and then finished on the wheel (Arnold, 1993, p. 36). Wodzińska (2009c, p. 25) suggests that the potter's wheel was used as a secondary means of production during Naqada II rather than as the primary means, which was coiling, pinching or slab construction.

Since Arnold (1993, pg 41-9), Holthoer (1977, pg 6-26) and Odler (forthcoming) have already discussed the various palaeographic and iconographic sources for the potter's wheel, and these sources have been outlined in Chapter 2, there is no need to go into further detail here. To date, the use of such secondary evidence has proved inconclusive and scholars have been unable to decide precisely when the potter's wheel began to be in use in Egypt, however, most consider the potter's wheel to have been in use by the 5th or 6th dynasty. Moreover, it must be stated that such secondary evidence cannot necessarily be viewed as verification for the use of the potter's wheel for throwing pottery; only the manufacturing marks on pottery can provide this evidence. Given that there is not sufficient pictorial evidence of the potter's wheel nor in the physical remains of potter's wheels prior to the 5th dynasty, there is a need to turn to the pottery itself to gain more objective evidence for its use.

The techniques involved in coiling and throwing pottery

In order to try to distinguish between these somewhat confusing pottery manufacturing terms, it would be prudent to address the issue of precisely what is meant by a coil made pot and what is a wheel thrown pot. A coil made pot comprises coils of clay which are formed by squeezing or rolling the clay into ropes whose diameter is usually two-three times the intended thickness of the vessel. Coil pots are usually built up by placing these ropes of relatively dry clay in a spiral formation and then smoothing down the sides. The joins of the coils can be difficult to discern and can lead ceramicists to think that the pot may have been thrown, particularly if the potter has smoothed down the clay by burnishing or has used a wash or slip to cover over the joins (Franken 2005, p. 14). This is regularly the case in Early Dynastic and Old Kingdom pottery which are frequently labelled "wheel finished" or "wheel rim rotated" e.g. black burnished or polished red wares (Petrie 1921, pp. IX-XIV, XIX), or Meidum bowls (Hendrickx *et al.* 2002, pp. 277-304).

By contrast, a wheel thrown pot is made entirely on a potter's wheel and is shaped by the potter's hand lifting the clay, aided by centrifugal force (Rye 1981, p. 74). When throwing, the potter uses clay which is softer and damper than that which is required in hand building so that the shape of the vessel can be easily drawn out and shaped. This is also done to try to negate the water evaporation caused by the air circulation during the rotation of the potter's wheel. The techniques involved in using a potter's wheel are entirely different to that of hand-building and require a stable forearm, the ability to be ambidextrous and the skill of knowing how much pressure to exert when throwing, depending on the plasticity of the clay, the speed of the wheel and the shaping method. This can only be achieved through experience and continuous practice and cannot be taught orally as it relies upon the potter learning how to position their body, arms, and hands precisely and firmly in order to achieve an accurately centred and thrown vessel (Birks, 1979; Cardew, 2002; Rado, 1969). As a result, throwing may take a long time to learn, possibly up to ten years (Ericson and Lehman 1996), whereas coil-built pottery may be mastered in two years (Roux 2003, p. 15). Coiling and other pottery hand manufactured techniques, although difficult in their own ways, can at least be attempted by all skill levels.

When throwing, the potter has to use a variety of highly mechanised movements (see video at http://bit.do/potterswheel). First, (though not in all cases) the clay is thoroughly wedged to remove air bubbles and impurities. It is then formed into cones or balls to render the centring process on the wheel easier. Once this is accomplished, the cone of clay is dropped/slammed on to the wheelhead so that it will stick and not slip off when the wheel is spun. The wheel is spun as fast as possible and the clay is centred on the wheel to reduce oscillation and allow the vessel wall and rim to be even. Next, the clay is opened out using the fingers and the potter begins to lift the vessel walls using thumb and forefinger. After which the vessel walls are shaped, trimmed with wooden ribs or other tools and the rim of the vessel is created. Once the potter is satisfied with their vessel, the pot is cut off the lump of clay with a piece of string or wire and left to dry. When green or "leatherhard", the vessel can be further trimmed or shaped up by being upturned on to the wheel, placed in a chuck or have handles added. It is then left to dry until hard enough to withstand a kiln firing (Cardew 2002, pp. 104-125; Leach, 1945, pp. 70-83, Rado, 1969; Rice 1987, pp. 128-9).

Experiment 1: Comparing the handbuilt coil and thrown pottery

To compare these two construction methods, an experiment was devised in which the author would make a series of thrown pots using an electric wheel and hand-built coil pots. These pots were made[34] using buff stoneware clay with the addition of iron oxide spangles so that they could later be x-rayed after firing. The pots were all fired in the same firing in an electric kiln. The two construction methods were filmed and photographed (see Figure 6.1), in order to deconstruct the gestures and movements made during manufacture and ascertain whether the techniques used could be associated with particular manufacturing marks produced on the pots. These marks would then be compared to archaeological pottery collections in museums. The pots were then broken and exposed to X-ray 40, 50 and 60 KV for two minutes each to ascertain whether further features would be revealed.

X-rays can provide further insight into the wheel thrown or hand-made origins of a pot which the naked eye or low magnification cannot identify. Frequently, pots that have been made using the coil technique can display horizontal lines similar to throwing. In addition, the final shaping of the pot can cover any marks formed during the original manufacturing processes (van der Leeuw 1976, p. 123); and sometimes the ascending spiral striations can be entirely due to hand building. Vandiver and Lacovara (1985) used xeroxradiography[35] to examine the porosity of the vessels and the alignment of the pores in order to try to identify the methods used for manufacture. They found that evenly spaced horizontal rows of horizontal-shaped pores parallel to the wall of the pot were thought to indicate coiling, whereas an even distribution of pores elongated in a diagonal direction (c 30° angle) indicated throwing (see Figure 6.2 and Figure 6.3). Pores randomly orientated may indicate slab manufacture. When the wheel thrown and coil-built experimental pots were examined in detail using x-rays, the differences detailed above were quite clearly evidenced (see Figure 6.2).

Even with the naked eye, the coils of the coilmade pot were quite easily seen. However, when examined under an X-ray, the coils become even more obvious, which would be exceedingly useful if the coils were not clear without a visual aid such as an X-ray. Additional details can be noted, the fingerprints of the potter are revealed, the join lines where the coils overlapped are more obvious and the smoothing lines and areas of depression where the fingers of the potter pushed into the clay are apparent. Under X-ray, the electric wheel thrown pot is entirely different to the coiled. Unlike the coiled pot, the wheel thrown pot has no join lines, no areas of overlapping and no areas of depression or fingerprint marks pushing into the clay. It has characteristic striations or rilling marks running continuously perpendicularly to the base of the pot, sticky finger marks are visible and a raised dimple or bump on the inside of the vessel. There are "drag" marks on the base of the vessel where a piece of wire or string was used to cut

---

[34] Valentine Stoneware clay V9A 1140-1280°C Buff clay, CTM Potters Supplies Ltd, used at Clayhill Pottery, Newnham on Severn, Gloucestershire, UK under supervision of professional potter Joan Doherty.
[35] An electrostatic technique giving an image enhancement due to a build up of charge at edges or density gradients.

FIGURE 6.1: THE WHEEL-THROWN POT AND THE HAND-BUILT COIL POT AT LEATHER HARD STAGE. PHOTO: S. DOHERTY

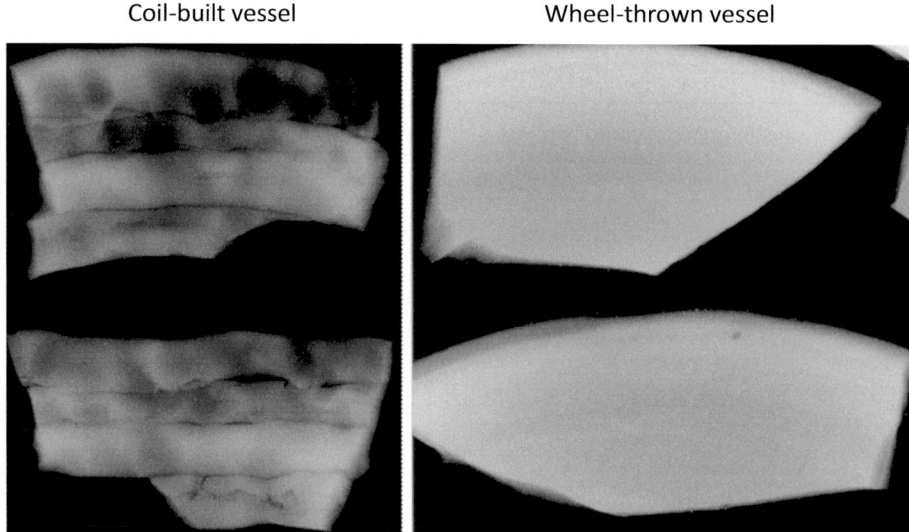

FIGURE 6.2: X-RAYS OF THE COIL HAND-BUILT EXPERIMENTAL POT (LEFT) AND ELECTRIC WHEEL-THROWN POT (RIGHT). THE WHITE "SPECKS" ARE IRON OXIDE SPANGLES THAT HAVE BEEN ADDED TO THE BUFF STONEWARE CLAY TO REPLICATE THE DISTRIBUTION OF ADDED TEMPER WITHIN THE CLAY. X-RAY: J. PEAKE AND S. DOHERTY AT CARDIFF UNIVERSITY, BOTH AT 40 KV FOR 2 MINUTES

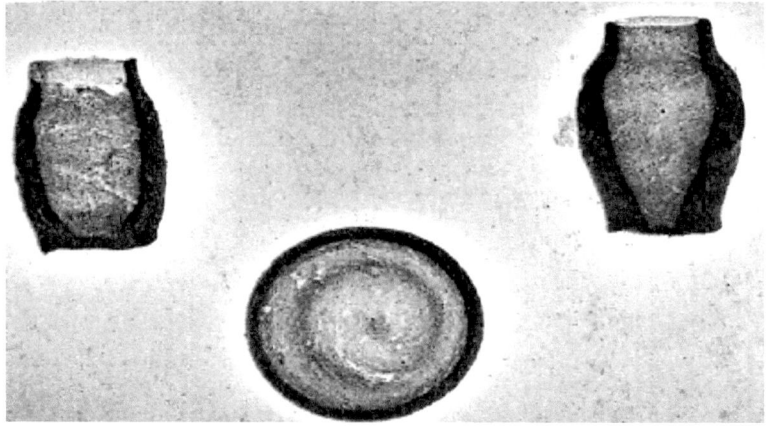

FIGURE 6.3: XERORADIOGRAPH OF THREE MINIATURE VESSELS. NOTE SPIRAL PATTERN IN THE BOWL (CENTRE) AND THE CROSS HATCHING IN THE WALLS OF THE JARS ON EITHER SIDE. EXPOSURE 150 kV, 18mAs AFTER: MAGRILL AND MIDDLETON 1997, PG 73, FIG 6(D)

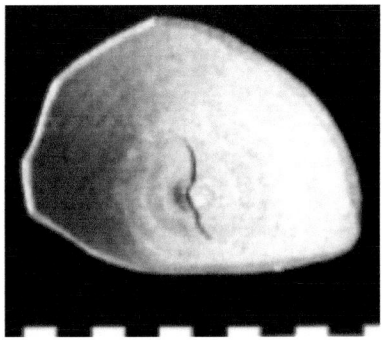

FIGURE 6.4: INDICATIONS OF THROWN POTTERY. LEFT: AN EXAMPLE OF AN S-SHAPED CRACK, INDICATIVE OF THROWN POTTERY, FROM GOBLET P03-219, TELL SABI ABYAD, SYRIA, LATE BRONZE AGE (DUISTERMAAT, 2008, PP. 379, FIG V27). RIGHT: THE INDICATIVE STRING CUT "DRAG" LINES LEFT BEHIND WHEN A POT HAS BEEN REMOVED FROM THE HUMP OF CLAY AFTER THROWING. PHOTO: S. DOHERTY

it from the clay on the wheel, and the rim is evenly formed. Other marks indicative of wheel-throwing can include a string cut base, a deep torsion crack or s-shaped crack on the inside of the base of the vessel (see Figure 6.4). These were not visible on the X-rays of the experimental pot as it had a raised dimple on the inside of the base rather than a crack (see Figure 6.10 and Figure 6.23), and the vessel was cut off the wheel using a wire when the wheel was stopped, so drag rather than string marks were visible.

Iron oxide spangles were added to the clay to try to reflect the orientation of temper. The iron oxide spangles do appear to demonstrate that in the coil made pots, the orientation of temper is random, whereas in the wheel thrown pot, the spangles align roughly parallel to the wall of the vessel. However, as the spangles were less than 5mm in diameter, they possibly were not large enough to demonstrate orientation significantly. If they were diagonally shaped rather than circular, the orientation of added temper to clay would perhaps be observable. Further experiments with larger sized iron oxide spangles would improve upon these results (see Figure 6.5).

FIGURE 6.5: IRON OXIDE SPANGLES BEING ADDED TO THE CLAY DURING THE WEDGING PROCESS. PHOTO: S. DOHERTY

As an alternative to X-rays and xeroxradiography, Courty and Roux (1995) suggested that if the grooves (the dips between the ridges along the interior) and rilling marks (spiral ridges or striations around the interior/exterior formed by finger pressures) were examined, it could be determined how to identify these in wheel thrown and hand-made vessels through experimental archaeology (see Figure 6.3). Grooves made during wheel manufacturing are made because of an impurity being dragged along (sharp or angular edges) or by fingernails (rounded edges). In contrast, the grooves made because of coils have rounded irregular edges with crossing ripples of compression which often narrow the neck or the base of the vessel and fissures on the grooves due to coils of different thickness being applied and not sufficiently smoothed. These coil characteristics were unable to be replicated on the wheel in the experimental archaeology tests conducted with modern-day Indian potters (Roux & Corbetta, 1989). Rillings made during wheel manufacturing are suggested by Courty and Roux (1995, p. 751-3) to be prominent bands with irregular edges. Rilling can however occur during wheel rotation, so correct diagnosis of manufacturing can be a problem. Other possible wheel manufacturing diagnostics could be cracks, particularly at the inner base of vessels, which seem to occur due to the high water usage when ceramics are thrown rapidly. It seems therefore, that a combination of application of X-rays, analysis of ancient pottery collections, and experimental recreation of ancient pottery is the best solution to the problem of identifying wheelthrown vessels.

Examination of museum collections

The next step was to examine the pottery collections of various museums to identify potentially wheel thrown pottery using the characteristics of wheel throwing and coil-building which had been classified in Experiment 1. The collections studied included the Ashmolean Museum, the Petrie Museum of Egyptian Archaeology, the British Museum, Cyfarthfa Castle Collections and the Egyptian Museum, Cairo. Some of the Predynastic and Early

Dynastic pots from these assemblages seemed to display certain characteristics associated with wheelthrown pottery. For example, many wavy handled jars have concentric rilling marks on the upper inside part of the vessel, usually to the depth of a finger span, making them appear to be thrown (see Figure 6.7 and Figure 6.8). However, upon further investigation this proved to not be the case (see Table 6.1).

Such vessels often have coils visible in the lower section of the vessel, which sometimes look like the characteristic rilling marks indicating the use of a potter's wheel, when in fact the potter's wheel has not been used (see Figure 6.6). Throwing the rim separately and adding it to the top of the vessel is an unlikely scenario, as it would be difficult to throw a vessel the same size as an uneven coil hand-built one, and it would probably take too much time. Many of these processes can be reflected upon the finished pot. Often, archaeological ceramicists focus on the rim and the neck of a vessel, rather than the base or the body of the vessel to identify how the pot was made and to identify its type. However, these results can be misleading, as potters sometimes rotated the neck of a pot which was otherwise hand-made. The rim of the vessel is continuously manipulated throughout the shaping process and is usually the last thing to be finished. On a completely thrown pot, one would expect the entire body of the pot to exhibit such characteristic marks as:

1. an S-shaped torsion crack or outward spiral at the base of the pot, reflecting stresses imposed during the opening of the vessel, sometimes also causing slumping if clay is not originally centred;[36]

2. diagonal orientation of voids and inclusions within the clay, some parallel to the work surface in cross section;

3. continuous, evenly distributed rilling marks with spiral grooves towards the inner centre;

4. a string cut base, or evidence for the vessel's removal from a lump of clay;

5. evidence for an exceptionally wet clay, such as sponge or cloth marks, as wheel throwing requires clay to be continuously wetted which coiling and other building techniques do not;

6. pots sometimes fracture in spiral shape. Fractures near the base suggest pressures imposed on this area by lifting.

However, upon closer inspection, the base of the pot revealed the hand-made origins of the vessel. The bases of wavy handled jars are flat without the string cutting marks associated with wheel thrown pots, and do not have s-shaped spirals or torsion cracks. The profiles of the wavy handled vessels are often uneven. This corresponds to the thinning and thickening of the coils of clay as they were pressed together during the forming of the vessel (see Figure 6.6). The rim of the vessel however, often displays

[36] Similar concentric lines may be visible on some coil made vessels, particularly around the rim, but no S-shaped cracks are visible.

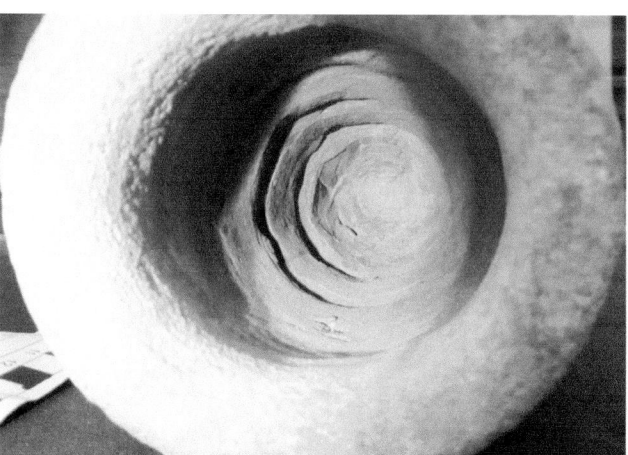

FIGURE 6.6: COILS CLEARLY VISIBLE IN THE BASE OF THIS WAVY HANDLED JAR C.3200 B.C. © ASHMOLEAN 546-95. PHOTO: S. DOHERTY

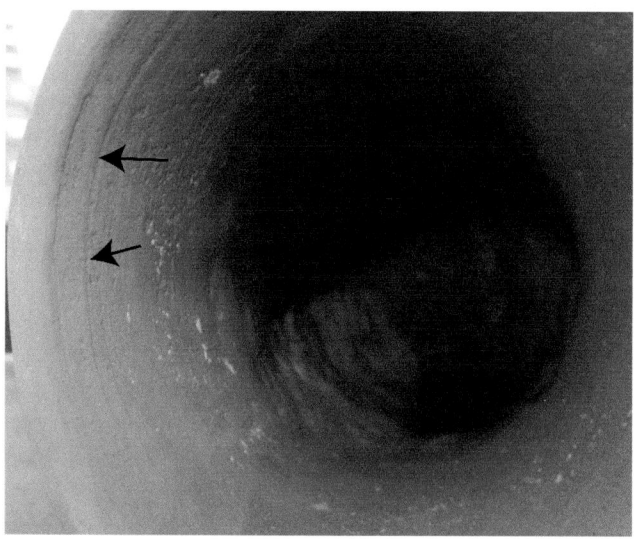

FIGURE 6.7: THE RILLING MARKS CREATED BY THE FINGERS OF THE POTTER (INDICATED BY ARROWS) WHEN SHAPING THE RIM OF THE WAVY HANDLED JAR. THE VESSEL WAS ENTIRELY HAND BUILT USING LARGE COILS OF CLAY, THEN PLACED WITHIN A STATIONARY CHUCK, MAT, OR SUPPORT AND THE ENTIRE POT SLOWLY ROTATED BY THE POTTER. RIM HEIGHT C5CM, WAVY HANDLED JAR C.3200 B.C. CYFARTHFA CASTLE COLLECTION 297.004. PHOTO: S. DOHERTY

concentric rilling marks similar to wheel thrown pottery (see Figure 6.7). Nevertheless, the vessel is not likely to have been rotated at speed as reasonably clear finger widths can often be detected (see Figure 6.7). The clay was also likely to have been damp rather than wet as the fingermarks are easily discernible. When throwing, the potter requires quite a wet clay to reduce friction. Instead of being thrown, it is likely that the pot was placed on a stand or chuck and coils built upon it while rotating the stand or pot, a technique still used by modern potters when

Museum	Museum No.	Date	Type	Provenance	Fabric	Rilling	Base	Wet	Primary method	S-shape	Other	Picture
Ashmolean	1895.567	Naq IId2	necked jar	Naqada 693		Rim	Flat		Coil		wavy	
Ashmolean	E.3653	Naq IId2	necked jar	el Amra b224		Rim	Flat		Coil		wavy	
Ashmolean	1932.912	Naq IIc	necked jar	Matmar 5115		Rim	Flat		Coil		tubular	
Ashmolean	1895.566	Naq IId1	necked jar	Naqada B101		Rim	Flat		Coil		wavy	
Ashmolean	1895.589	Naq IId1	necked jar	Naqada 1686		Rim	Flat		Coil		tubular	
Ashmolean	1895.616	Naq IId2	necked jar	Naqada 818		Rim	Flat		Coil		-	
Ashmolean	1895.586	Naq IIb	necked jar	Naqada 1639			Flat		Coil		tubular	
Ashmolean	1924.328	Naq IIc	necked jar	Badari 4602			Flat		Coil		tubular	
Ashmolean	E.2822	Naq IIc	necked jar	Hu U136			Flat		Coil		tubular	
Ashmolean	1891.556	Naq IIc	necked jar	Abydos			Flat		Coil		tubular	
Ashmolean	1892.1074	Naq IIc	necked jar	Sigareieh			Flat		Coil		tubular	
Ashmolean	1895.574	Naq IIb	necked jar	Naqada 1729			Flat		Coil		tubular	
Ashmolean	1892.1075	Naq IIc	necked jar	Sigareieh			Flat		Coil		tubular	
Ashmolean	1895.608	Naq IId1	necked jar	Naqada 625		Rim	Flat		Coil		tubular	
Ashmolean	1961.301	Naq IId1	necked jar	unknown		Rim	Flat		Coil		tubular	
Ashmolean	1895.569	Naq IId1	necked jar	Naqada 665		Rim	Flat		Coil		tubular	
Ashmolean	1935.112	Naq IId1	necked jar	Armant 1408		Rim	Flat		Coil		tubular	
Ashmolean	1891.22	Naq IId1	necked jar	Semaineh		Rim	Flat		Coil		tubular	
Ashmolean	1895.572	Naq IId1	necked jar	Naqada 625		Rim	Flat		Coil		tubular	
Ashmolean	1891.21	Naq IId1	necked jar	Semaineh		Rim	Flat		Coil		tubular	
Ashmolean	E.2821	Naq IId1	necked jar	Abadiya B360		Rim	Flat		Coil		tubular	
Ashmolean	1895.597	Naq IIb	necked jar	Naqada 1766			Flat		Coil		tubular	
Ashmolean	E.2880	Naq IId2	double necked jar	Semaineh H5			Rounded		Coil		lug, vertically pierced	
Ashmolean	1933.275	Naq IId1	necked jar	Khozam			Flat		Coil		tubular	
Ashmolean	E.2803	Naq II	necked jar	Abadiya B			Flat		Coil		tubular	
Ashmolean	1891.2	Naq II	necked jar	Semaineh			Flat		Coil		tubular	
Ashmolean	E.2881	Naq IId2	necked jar	Semaineh H39?			Flat		Coil		-	
Ashmolean	1895.588	Naq IId1	necked jar	Naqada 100?		Rim	Flat		Coil		tubular	
Ashmolean	1892.1073	Naq IIb	necked jar	Semaineh			Flat		Coil		tubular	
Ashmolean	1895.587	Naq IIb	necked jar	Naqada 1657			Flat		Coil		tubular	
Ashmolean	Queen's Coll 1248	Naq IIc	necked jar	Armant 1363			Flat		Coil		tubular	
Ashmolean	1891.19	Naq IIc	necked jar	Semaineh			Flat		Coil		tubular	
Ashmolean	1891.23	Naq IIc	necked jar	Semaineh			Flat		Coil		tubular	

TABLE 6.1: MANUFACTURING MARKS CRITERION

The Origins and Use of the Potter's Wheel in Ancient Egypt

Museum	Museum No.	Date	Type	Provenance	Fabric	Rilling	Base	Wet	Primary method	S-shape	Other	Picture
Ashmolean	1964.304	Naq IId1	double necked jar	unknown			Flat		Coil		tubular	
Ashmolean	1895.573	Naq IIc	necked jar	Naqada 1740			Flat		Coil		tubular	
Ashmolean	1966.356	Naq IId1	necked jar	unknown			Flat		Coil		tubular	
Ashmolean	1895.585	Naq IId1	necked jar	Naqada 409			Flat		Coil		tubular	
Ashmolean	1891.18	Naq IId1	necked jar	Semaineh			Flat		Coil		tubular	
Ashmolean	1895.815	Naq IIc	necked jar	Naqada south			Flat		Coil		tubular	
Ashmolean	E.2877	Naq IIc	necked jar	Hu U122			Flat		Coil		tubular	
Ashmolean	1933.1415	Naq IIc	necked jar	unknown			Flat		Coil		tubular	
Ashmolean	E.2878	Naq IIc	necked jar	Hu U203			Flat		Coil		tubular	
Ashmolean	1895.571	Naq IIc	necked jar	Naqada 1852			Flat		Coil		tubular	
Ashmolean	1933.1416	Naq II c	necked jar	unknown			Flat		Coil		tubular	
Ashmolean	1895.606	Naq IId2	necked jar	Naqada 173			Flat		Coil		tubular	
Ashmolean	1895.595	Naq IId2	necked jar	Naqada 1268			Flat		Coil		tubular	
Ashmolean	1948.17	Naq IId2	necked jar	Aswan			Flat		Coil		tubular	
Ashmolean	1955.566	Naq IId1	necked jar	Thebes			Flat		Coil		tubular	
Ashmolean	1895.577	Naq IId1	simple jar	Naqada 1873			Rounded		Coil		lug, horizontally pierced	
Ashmolean	1895.578	Naq IId1	simple jar	Naqada 1680			Rounded		Coil		lug pierced	
Ashmolean	1933.845	Naq IId1	simple jar	Upper Egypt			Flat		Coil		lug pierced	
Ashmolean	1895.6	Naq IId2	necked jar	Naqada 1209			Flat		Coil		-	
Ashmolean	1895.598	Naq IId2	necked jar	Naqada 1458		Rim	Flat		Coil		tubular	
Ashmolean	1895.593	Naq IId1	simple jar	Naqada 690			Flat		Coil		lug pierced	
Ashmolean	E3968	Naq IId2	necked jar	Hu U177			Flat		Coil		tubular	
Ashmolean	1966.537	Naq IId2	necked jar	unknown			Flat		Coil		tubular	
Ashmolean	1891.24	Naq IId2	necked jar	Semaineh		Rim	Flat		Coil		tubular	
Ashmolean	E2824	Naq IId2	necked jar	Hu U128			Flat		Coil		tubular	
Ashmolean	1933.142	Naq IId2	necked jar	unknown			Flat		Coil		-	
Ashmolean	E 2876	Naq IId2	necked jar	Semaineh H8			Flat		Coil		inner ledge pierced	
Ashmolean	1895.612	Naq IId1	necked jar	Naqada 643			Flat		Coil		-	
Ashmolean	1895.1235	Naq IId1	necked jar	Naqada 562			Flat		Coil		tubular	
Petrie	UC20082	Old Kingdom	Bottle	Buhen	NB1	Body	Unknown	yes	Wheel-thrown			101-0079, 81
Petrie	UC20083	Old Kingdom	Jar	Buhen	NB1	Body & Rim	Unknown	yes	Wheel-thrown			
Petrie	UC20084	Old Kingdom	Dish?	Buhen	NB1	Rim (sherd)	Unknown		Wheel-thrown			101-0082
Petrie	UC20085	Old Kingdom	Medum ware	Buhen	NB1	Rim (sherd)	Unknown		Wheel-thrown			
Petrie	UC20086	Old Kingdom	Jar	Buhen	NB1	Rim (sherd)			Wheel-thrown		incurved narrow neck	101-0084 -0087

Detecting the Use of the Potter's Wheel on Ancient Egyptian Pottery

Museum	Museum No.	Date	Type	Provenance	Fabric	Rilling	Base	Wet	Primary method	S-shape	Other	Picture
Petrie	UC20087	Old Kingdom	Medum ware	Buhen	NB1	Body	Rounded		Wheel-thrown			
Petrie	UC20088	Old Kingdom	Medum ware	Buhen	NB1	Body	Rounded		Wheel-thrown			101-0090-1
Petrie	UC20089	Old Kingdom	Medum ware	Buhen	NB1	Body			Wheel-thrown			
Petrie	UC20090	Old Kingdom	Medum ware	Buhen	NB1	Body			Wheel-thrown			101-0093
Petrie	UC20091	Old Kingdom	Medum ware	Buhen	NB1	Body			Wheel-throw			101-0094
Petrie	UC20092	Old Kingdom	Jar	Buhen	NB1	Body			Wheel-thrown			101-0096,98
Petrie	UC20093	Old Kingdom	Medum ware	Buhen	NB1	Body			Wheel-thrown		small	101-0099, 100
Petrie	UC20094	Old Kingdom	Medum ware	Buhen	NB1	Body			Wheel-thrown			
Petrie	UC20095	Old Kingdom	Medum ware	Buhen	NB1	Body			Wheel-thrown			101-0101-102
Petrie	UC20096	Old Kingdom	Medum ware	Buhen	NB1	Body			Wheel-thrown			
Petrie	UC20097	Old Kingdom	Medum ware	Buhen	NB1	Body			Wheel-thrown			
Petrie	UC20098	Old Kingdom	Medum ware	Buhen	NB1	Body			Wheel-thrown			101-0104,5
Petrie	UC20099	Old Kingdom	Medum ware	Buhen	NB1	Body			Wheel-thrown			101-0106
Petrie	UC20100	Old Kingdom	Medum ware	Buhen	NB1	Body			Wheel-thrown			101-0107,8
Petrie	UC20101	Old Kingdom	Medum ware	Buhen	NB1	Body			Wheel-thrown			101-0109
Ashmolean	E550.1901	Old Kingdom	Jar	Bet Khallim		Rim	Scraped		Coil			81
Ashmolean	1935-110	Naq IIIa1	Jar		Marl	Rim	Scraped		Coil		red paint waves	88,89,91
Ashmolean	E1065.93	Naq IIId	Jar		Marl		Flat		Coil		wavy handled	92,93,94
Ashmolean	1895.766	Old Kingdom	Miniature jar		NB1	Body	String cut	yes	Wheel-thrown	yes		105, 114
Ashmolean	E546.95	Naq IIc	jar		NB1	Rim	Flat		Coil		wavy handled	120-123
Ashmolean	E555.95	Naq IIid	Jar		Marl	Rim	Flat		Coil		corded	130, 131, 132,
Ashmolean	E1895.719	Naq III	Jar		NA	Rim	Rounded		Coil		string impressed	138
Ashmolean	E531.95	Naq IIIc	Jar		NB	Rim	Flat		Coil		wavy handled	139
Ashmolean	E543.95	Naq IIc	Jar		NB	Rim	Flat		Coil		wavy handled, black paint	140, 143
Ashmolean	547.95	Naq IIId	Jar		Marl	Rim	Flat		Coil		wavy handled	146, 148
Ashmolean	E3654	Naq IId	Jar	Hu	Marl	Rim	Flat		Coil		D-ware, red scorpions	152
Ashmolean	E1892.1060	Naq II	Jar		NA		Rounded		Coil		D-ware, red comma	153, 159
Ashmolean	E2825	Naq IId	Jar		Marl		Flat		Coil		D-ware,bag shaped	160
Ashmolean	E 546.95	Naq IIId	Jar		Marl	Rim	Flat		Coil		Wavy handled	164, 165, 167
Ashmolean	E515.1901	Naq II	Bowl		NA		Flat		Coil			168
Ashmolean	1891. E588	Naq IIIc1	Bowl		NA		Flat		Mould		scraped	176
Ashmolean	1891.E586	Old Kingdom	Medum ware		NB1		Flat		Mould			179
Ashmolean	1902.E498	Old Kingdom	Medum ware	Regarah	NB1	Body	Rounded	yes	Wheel-thrown	yes	Carinated	183, 181
Petrie	UC17624	Old Kingdom	Miniature jar	Medum	NB1	Body	String cut	yes	Wheel-thrown	yes	4th dynasty	
Petrie	UC17625	Old Kingdom	Miniature jar	Medum	NB1	Body	String cut	yes	Wheel-thrown	yes	4th dynasty	
Petrie	UC17626	Old Kingdom	Miniature jar	Medum	NB1	Body	String cut	yes	Wheel-thrown	yes	4th dynasty, mastaba 18	

Museum	Museum No.	Date	Type	Provenance	Fabric	Rilling	Base	Wet	Primary method	S-shape	Other	Picture
Petrie	UC17608	Old Kingdom	Miniature jar	Medum	NB1	Body	String cut	yes	Wheel-thrown	yes	4th dynasty	
Petrie	UC17622	Old Kingdom	Medum ware	Medum	NB1	Body			Wheel-thrown		4th dynasty, chalice	
Petrie	UC18404	Old Kingdom	Miniature jar		NB1	Body	String cut	yes	Wheel-thrown	yes	series of jars on stand	
Petrie	UC17366	Old Kingdom	Miniature jar	Abydos	NB1	Body	String cut	yes	Wheel-thrown	yes	still on hump of clay	
Petrie	UC17617	Old Kingdom	Miniature Cup	Medum	NB1	Body	String cut	yes	Wheel-thrown	yes	polished, Sneferu foundation	
Cairo	51852	Old Kingdom	Miniature Cup			Body	String cut	yes	Wheel-thrown	yes	room 48, case E	
Cairo	49254	Old Kingdom	Miniature jar			Body	String cut	yes	Wheel-thrown	yes	room 48, case E	
Cairo	51851	Old Kingdom	Miniature Cup			Body	String cut	yes	Wheel-thrown	yes	room 48, case E	
Cairo	49255	Old Kingdom	Miniature jar			Body	String cut	yes	Wheel-thrown	yes	room 48, case E	
Cairo	49253	Old Kingdom	Miniature jar			Body	String cut	yes	Wheel-thrown		room 48, case E	
Cairo	49256	Old Kingdom	Miniature jar			Body			Wheel-thrown		room 48, case E	
Cairo	49244	Old Kingdom	Miniature jar			Body		yes	Wheel-thrown		room 48, case E	
Cairo	49338	Old Kingdom	Miniature jar			Body			Wheel-thrown		room 48, case E	
Cairo	49249	Old Kingdom	Miniature jar			Body	String cut	yes	Wheel-thrown	yes	room 48, case E	
Cairo	51818	Old Kingdom	Miniature jar			Body	String cut	yes	Wheel-thrown	yes	room 48, case E	
Cairo	51819	Old Kingdom	Miniature jar			Body	String cut	yes	Wheel-thrown	yes	room 48, case E	
Cairo	49249	Old Kingdom	Miniature jar			Body		yes	Wheel-thrown	yes	room 48, case E	
Petrie	UC17634	Old Kingdom	Miniature plate	Medum	NB1	Body	String cut	yes	Wheel-thrown	yes	4th dynasty, Sneferu foundation	
Petrie	UC17608	Old Kingdom	Miniature jar	Medum	NB1	Body	String cut	yes	Wheel-thrown	yes	4th dynasty, Sneferu foundation	
Petrie	UC17618	Old Kingdom	Bowl	Medum	NB	Body	String cut	yes	Wheel-thrown	yes	4th dynasty, Sneferu foundation	
Cyfarthfa	Cy277.004	Naq II	Jar		Marl				Coil		red paint boats	0139
Cyfarthfa	Cy232.004	Naq III	Jar		NB1	Rim			Coil		wavy handled cord	0142, 0143, 0144
Cyfarthfa	Cy297.004	Naq III	Jar	Tarkhan	NB1	Rim			Coil		red paint basket, wavy handled	0145, 0146
Cyfarthfa	Cy280.004	Naq III	Jar	Tarkhan	M A1	Rim			Coil		red paint basket, wavy handled	0151, 0148, 0145
Cyfarthfa	Cy270.004	Naq II	Jar		Marl						triangles, ferns, red paint	159
Petrie	UC17301	Naq IIId	Jar	Tarkhan	Marl	Rim			Coil		Wavy handled, red basket	219
Petrie	UC17208	Naq IIId	Jar	Tarkhan	Marl	Rim			Coil		Wavy handled	
Petrie	UC17545	Naq IIIc	Jar	Hierakonpolis	Marl	Rim			Coil		Wavy handled, corded	197
Petrie	UC17287	Naq IIIc	Jar	Tarkhan	NB	Rim			Coil		Wavy handle	203, 4
Petrie	UC17297	Naq IIIc	Jar	Tarkhan	NB	Rim			Coil		Wavy handled, red basket	208, 210
Petrie	UC17445	Naq IIId	Jar	Abydos	NB1	Rim			Coil		Wavy handled, corded	
Petrie	UC17345	1st Dynasty	Necked jar	Abydos	NB	Rim			Coil		Rolled rim	235, 231

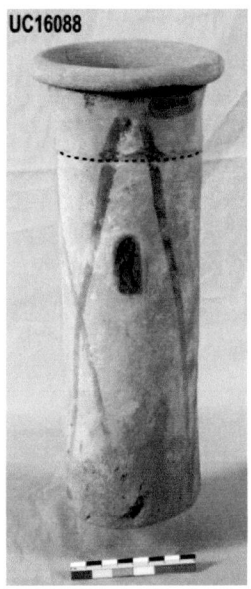

FIGURE 6.8: WAVY HANDLED JAR. CONSTRUCTED USING COILS ON FLAT SUPPORT (NOTE BASE OF POT IS FLAT), THEN A SLAB OF CLAY ADDED TO THE TOP OF POT AT JOIN LINE (INDICATED BY DOTTED LINE) WHICH WAS THEN ROTATED LEAVING RILLING MARKS ONLY IN INSIDE THE RIM AREA WHERE THE POTTER'S FINGERS WORKED TO SHAPE AND SUPPORT THE VESSEL. ©PETRIE MUSEUM OF EGYPTIAN ARCHAEOLOGY, UNIVERSITY COLLEGE LONDON UC16088

FIGURE 6.9: MINIATURE VESSEL FROM ABYDOS. THROWN ON A POTTER'S WHEEL. NOTE THE STRIATION MARKS, THE SCRAPES ON THE BASE AND RIM, AND STICKY FINGERMARKS. THE RIM HAS BEEN CAREFULLY SHAPED, BUT THE POTTER LEFT QUITE A LOT OF CLAY ON THE BASE WHEN CUTTING IT FROM THE WHEEL. HEIGHT 10.6CMS ©PETRIE MUSEUM OF EGYPTIAN ARCHAEOLOGY, UNIVERSITY COLLEGE LONDON UC17366

they use a "banding wheel" or stand. A final slab of clay was then placed on top (note the join line in Figure 6.8), the rim everted through slow rotation on the stand with the fingers on the inside of the vessel, leaving rilling lines. The addition of a slab allows any uneven coils to be easily smoothed and gives the rim of the pot a more even finish.

Wavy handled jars are regularly labelled "hand turned" implying that the vessel was rotated in some way, when the term "turned" to a potter indicates that the vessel was trimmed of excess clay.

Accordingly, it has been determined that Early Dynastic pottery such as the wavy handled jars do not possess wheel thrown characteristics, as the vessels do not display the wheel thrown manufacturing marks observable in the experimental vessels (see Table 6.2). Consequently, the museum pottery collections were once again consulted to verify whether there might be any wheel thrown pottery occurring before the 6th dynasty of the Old Kingdom. The 5th-6th dynasties are often the periods considered by Egyptologists to be when the potter's wheel comes into use in Egypt (Arnold, 1993, pg. 41; Senussi, 2006, pp. 329-330; Vachala & Faltings, 1995, pg. 282).

Miniature vessels

Amongst the pottery corpus of the early Old Kingdom there may be a candidate for wheel thrown vessel production, namely, the miniature vessel (see Figure 6.9); its potential has already been identified by Bárta (1995, pp. 15-24). These vessels were used as part of the daily offering rituals in chapels and pyramid temples in cults dedicated to the *ka* (Gahlin 2001) of a Pharaoh or private individual. Such mortuary cults became important during the time of monumental pyramid building by the Pharaohs of the 4th-5th dynasties (c.2600-2300 B.C.). As part of their funerary pyramid complexes, these Kings (beginning with Sneferu at Meidum, see Chapter 5) constructed cultic temples.

Sneferu conceived these new temples and specifically dedicated them to the nourishment of his *ka*[37] in the afterlife (Rowe 1931, pp. 14, 28-34). These rituals required daily offerings of food, drink and other items to be presented to a representation of the *ka* of the deceased king by a priest. Private individuals also had chapels to provide for their own *ka's* offerings. As these offerings occurred continuously, they required a regular supply of small vessels produced in large numbers for the food offerings or for symbolic offerings to the *ka*. Bourriau (1981, pp. 20, fig 11) suggests that these miniature vessels were entirely votive and never intended to be used as a container, apart from symbolically for the funerary cult.

---

[37] The *ka* is sometimes referred to as the spiritual double or vital force of the deceased person, and was intimately linked to the body as it served as the home for the *ka* after death. This was why the Egyptians practised mummification, to ensure that the *ka* had somewhere to reside after death. It was believed that the *ka* required food and drinks, so offerings were made to it and led to the rise of funerary cults (Gahlin, 2001).

The Origins and Use of the Potter's Wheel in Ancient Egypt

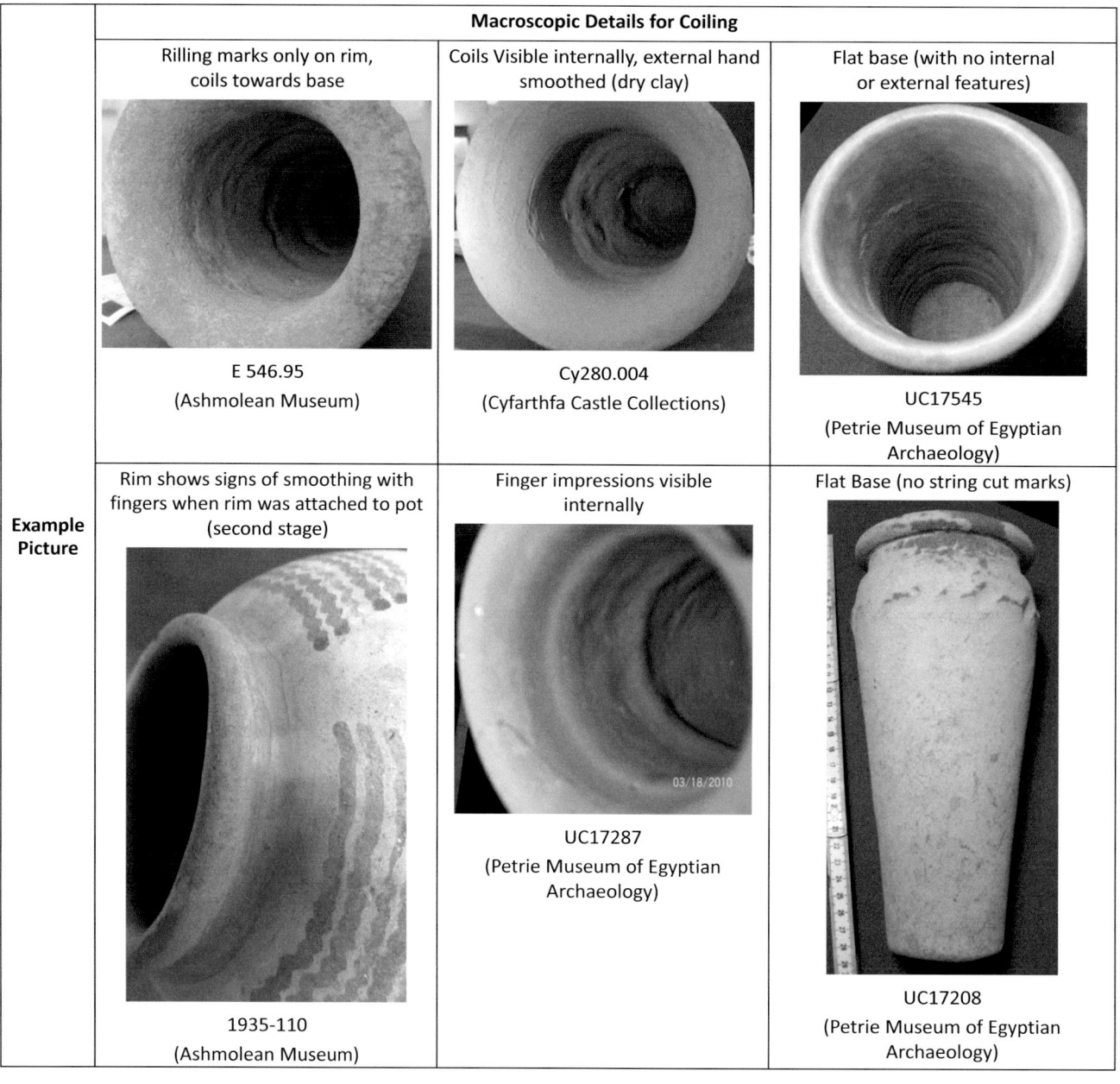

TABLE 6.2: EXAMPLES OF MACROSCOPIC DETAILS FOR COILING IN MUSEUM POTTERY COLLECTIONS

These vessels were, as their name suggests, very small, the largest being up to c7.5cms tall and c6.5cms wide. Both miniature vessels and model stone vessels were known from the Predynastic Period (Köhler 1992, pp. 7-8; these differences will be further discussed in Chapter 7), but miniature vessels only began to be wheel thrown during the reign of Pharaoh Sneferu, first king of the 4th dynasty (Bárta 1995, p. 15). They were mass-produced and designed to be used once, therefore many thousands are often excavated in pristine condition (e.g. some 10,000 were found at Dahshur by Fakhry 1961, p. 135). Unlike their contemporaries the similarly sized model vessels, miniature vessels have an interior volume, and could probably contain a token amount of liquid or grain and therefore retain their functional capability (Allen, 2006; D'Auria *et al.* 1988, pp. 77-78; Swain 1995). Miniature vessels are regularly discovered during excavations either as surface finds within burial chambers (Hassan 1948, p. 18), foundation deposits e.g. the pyramid of Sneferu at Meidum (Petrie *et al.* 1910, pp. 12, pl XXV nos 20-14; 32-14; Rowe 1931, pp. 28-30, pl XV; el-Khouli 1991, p. 13; Clayton 1994, p. 45; Dodson 1995, p. 27), in pits near to pyramid temples e.g. Menkaure's pyramid temple excavated by Reisner (1931, p. 228), funerary chapels or incorporated into the walls of funerary architecture, such as in the mastaba of Ptahshepses (Charvát 1981, p. 149).

Miniature vessels are often overlooked by excavators as they usually occur in great quantities and can be thought of as quite crudely fashioned; perhaps owing to the speed at which they were made. They are sometimes recorded in pottery reports thus:

Detecting the Use of the Potter's Wheel on Ancient Egyptian Pottery

	Macroscopic Details for Wheel-throwing		
Example Picture	Rilling marks throughout body of vessel  UC20082 (Petrie Museum)	S-shaped torsion crack or outward spiral at the base 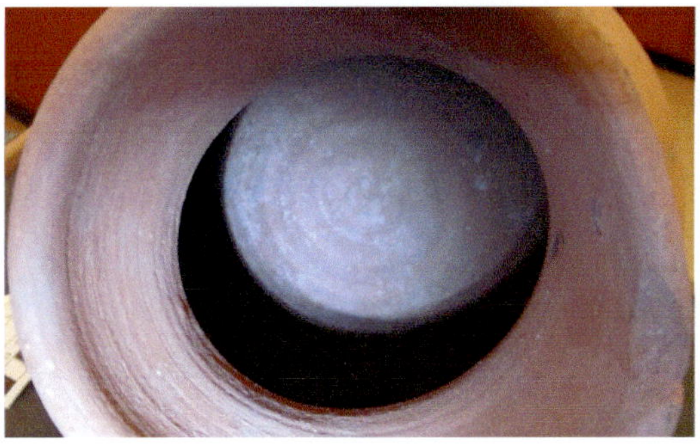 1895.719 (Ashmolean Collection)	String Cut Base  1895.766 (Ashmolean Collection)
	Fractures from base outwards  1902.E498 (Ashmolean Collection)	Sticky finger marks indicative of lots of water being used, "frilly" and uneven base  1895.766 (Ashmolean Collection)	

TABLE 6.3: TABLE OF MACROSCOPIC DETAILS FOR WHEEL THROWING IN MUSEUM COLLECTIONS

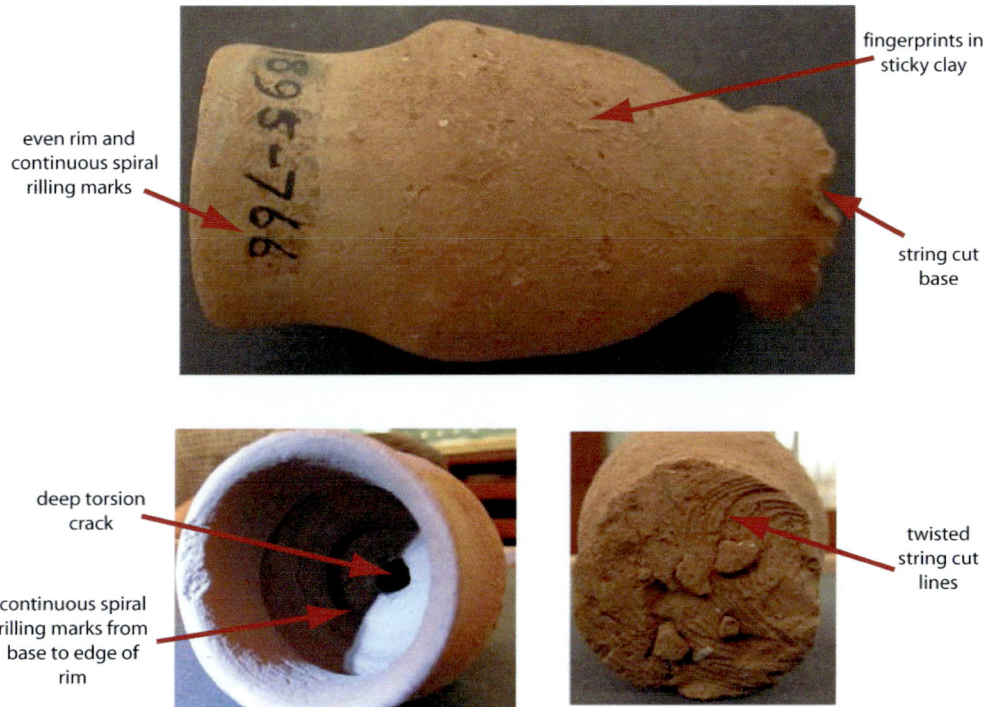

FIGURE 6.10: THE CHARACTERISTIC MARKS OF WHEEL-THROWING, AS INDICATED UPON THIS EXAMPLE OF A MINIATURE VESSEL. AN1895.766, ASHMOLEAN MUSEUM, OXFORD. PHOTOS: S. DOHERTY

"carelessly made, uneven, lopsided, some very warped, with uneven bases and large clumps of clay sticking to the surface...often with finger marks on the base,"(Milward Jones 1991, p. 45 forms 15 and 16).

It is speculated that such descriptions may represent the first wheel thrown vessels, perhaps the earliest known throughout the Near East, made by potters learning to utilise their new technology. Previously, Near Eastern potters were using the potter's wheel to finish coil-built pots rather than throwing. The miniature vessels are small, crude and lopsided, indicating that the potters had perhaps yet to master the intricacies of centring before forming the vessel. However, they also contain fine continuous striations and grooves, sticky finger marks on the sides of the vessel, s-shaped cracks, torsion wells, string cut bases and evidence of scraping (see Figure 6.10). Consequently, since these pots are prime candidates for providing evidence of early throwing in ancient Egypt, it was decided to try to replicate them using a reconstructed potter's wheel. With varying degrees of success, similar experiments have already been done to replicate ancient pottery from both Egypt and the Near East, and these have been outlined in Chapter 2.

Choosing a suitable potter's wheel

The majority of the potter's wheel bearings in the museum collections of Cairo, Oxford and London are comprised of an upper pivot and a lower socket stone usually of basalt, granodiorite or limestone. They range from 15cm-24cm in diameter and vary in height from 5.5-6cm (Powell 1995, pp. 309-311). Authors have suggested that 80 r.p.m. is sufficient to throw pots (Amiran and Shendov 1984; Rye 1981, p. 74). Modern potters using electric wheels suggest that a minimum speed of 50 r.p.m. and maximum of 130 r.p.m. are the optimum speeds for using an electric wheel

FIGURE 6.11: BM32622. THE POTTER'S WHEEL BEARINGS CHOSEN TO REPLICATE BY THE AUTHOR, COMPRISING A SOCKET OF BLACK GRANITE AND A PIVOT OF WHITE LIMESTONE. HERE THE PIVOT HAS BEEN PLACED INTO THE SOCKET, AS IT WOULD HAVE BEEN WHEN USED. SCALE IS 5CM. ©THE TRUSTEES OF THE BRITISH MUSEUM. PHOTO: S. DOHERTY

(Colbeck, 1982, p. 19). Therefore, it seems that provided a potter has the skill set, they should be able to achieve sufficient momentum by quickly rotating the wheel, and then be able to throw with two hands before the wheel slows again and needs to be rotated once more. However, Powell (1995) notes that it is difficult to throw pots of a larger size e.g. beer jars. Powell (1995) undertook experiments to test the optimum sort of lubricant to use (linseed) and the best size and shape of wheelhead for throwing clay (mixture of

FIGURE 6.12: BM32622. THE POTTER'S WHEEL BEARINGS CHOSEN TO REPLICATE BY THE AUTHOR, COMPRISING A SOCKET OF BLACK GRANITE AND A PIVOT OF WHITE LIMESTONE. SCALE IS 5CM. ©THE TRUSTEES OF THE BRITISH MUSEUM. PHOTO: S. DOHERTY.

FIGURE 6.13: THE NEWLY CURED CONCRETE POTTER'S WHEEL BEARINGS BASED ON THE BRITISH MUSEUM EXAMPLE BM32622. PHOTO: ALAN DAVIES

silt and desert sand) and concluded that a wheelhead with a diameter of 60cm is best (Powell, 1995, p. 323).

Given that much experimental work has already been undertaken, the author decided to follow similar methods to the above experiments and try to improve upon their results by undertaking a series of investigations into the use and understanding of the potter's wheel. Following Powell's (1995, p. 334) advice to use a smaller set of wheel bearings to replicate than the set that she used (BM32621), as Powell thought that the larger wheel bearings were too cumbersome to achieve sufficiently fast speeds it was decided to select BM32622 (see Figure 6.11 and Figure 6.12). This comprised of a lower socket stone of granite c18cm diameter socket, weighing 13.1kg, and an upper pivot stone of limestone c24cm in diameter. The pivot weighs 6.2kg. and the polished lower surface of the pivot measures 14.5cm across and 9.9cm high. The tenon measures 8.5cm in diameter at its base, with a height of 4.7 cm. When fitted together, the upper stone projects some 3-4cm from the lower stone. The socket stone is made from black granite, and measures 18cm in diameter and 10cm in height; the polished surface is 16cm, and the socket well 8 x 5.3cm deep. The base of the well and outer edge of the polished area are worn. The stone is neatly carved round with a flat base (see Figure 6.11).

Reconstructing the potter's wheel

The replica wheel bearings were initially made as prototypes in concrete and then later made in stone. Concrete is a cheap alternative to basalt or limestone but can still replicate some of the characteristics of stone. A concrete mix was specially devised by the School of Engineering at Cardiff University, UK. The plans based on BM32622 are used to construct the replica concrete and granite wheel bearings and are included in the Appendix IV. The manufacturing technicians, under the guidance of Prof. Alan Davies, designed and made moulds of thermosetting plastic, based on drawings and photos taken by the author of the examples in the British Museum.[38] The top and bottom moulds were then coated with PVA (Poly Vinyl Acetate) so that the concrete and water would not penetrate. Moulds were used so that the concrete could be cast completely, without the need for wire mesh reinforcements. Because the wheel bearings were made in a mould, the socket had to be formed hollow so that it could be removed (see Figure 6.13). Additional mortar had to be placed within the lower bearing so that it could be formed into a socket, the pivot could fit within it and lubricant be placed within the well. The concrete mix was left to form for 7 days and then placed in a curing tank filled with water for 28 days to achieve full strength. Once dried and cured, the replica wheel bearings had then to be "run in" initially using water and sandpaper so that any uneven surfaces could be smoothed. During these initial experiments the wheel bearings were found to move around the floor excessively and so a stand was needed to keep the bearings in place. The ancient potters were likely to have a permanent workshop with their wheel embedded into the ground, but unfortunately this was not possible in the university environment. A wooden stand with four bolts was made, so that the wheel bearings would be stable yet portable.[39]

Next, two wheel heads in the shape of circular discs were constructed; one was made of wood[40] and the other of fired clay.[41] Archaeologically, fired and unfired clay wheel heads

---

[38] The concrete mixture consisted of 53.6% Limestone Aggregate, 24.7% Sand, 13.4% Ordinary Portland Cement, 8.3% Water. The pivot weighed 6.96kg and the socket 5.68kg.
[39] The potter's wheel base stand was made by Steve P Meade, technician at the Mechanical Engineering Dept, ENGIN, Cardiff University.
[40] The wooden disk wheel head was made by the author using a deconstructed mahogany tabletop.
[41] The clay disk wheel head was made and fired at Clay Hill pottery, Cinderford, Gloucester by the author with the assistance of potter Joan Doherty.

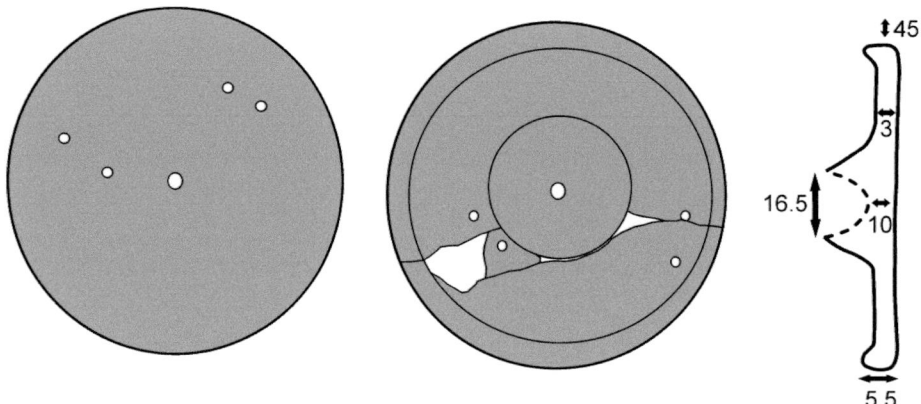

FIGURE 6.14: THE ORIGINAL SKETCH OF THE POTTER'S WHEELHEAD FOUND IN THE MORTUARY TEMPLE OF QUEEN KHENTKAUS II (AFTER THE FIND CARD OF THE EXCAV. NO. 293/A/78). THE NUMBERS ARE IN CENTIMETRES. ODLER (IN PRESS FIG 12)

are known from Ur (Simpson 1997b, pp. 50, fig 51), at Abu Sir (Verner 1992; 1995, pp. 27, fig 27a, pl 25) and various examples from Crete (Childe 1954, p. 201; Xanthoudides 1927), although most were fired. In tomb depictions e.g. Beni Hassan, the tomb of Gemniemhat (Nicholson and Doherty forthcoming; Holthoer 1977, p. 11; Newberry 1893) and Middle Kingdom and First Intermediate Period wooden models e.g. Inpuemhet (Quibell and Hayter 1927, pp. 40-41, pl 24), some wheelheads are painted a brown or reddish colour, which could indicate wood or fired clay. Grey colour usually indicated unfired clay. Powell (1995, pp. 322-324) suggests that an unfired heavily tempered clay disk could have been used; however, it is uncertain how durable it would be as the unfired wheelhead might mix with the clay to be thrown. Powell also experimented in using a wooden wheelhead, but determined that it was too light to be of use in throwing (Powell 1995, pp. 320, 322). During experiments on wheel heads of varying sizes, Powell (1995, pp. 330-332) determined that unfired wheel heads of c55cm were the most successful. Since Powell did not have the opportunity to try a heavier or larger wooden wheel head, it was thought useful to include this in the experiments.

Experiment 2: throwing on the replica potter's wheel

The author made a wooden wheelhead from an old table and constructed a fired clay wheelhead c55cm in diameter[41] (similar to Powell's (1995) optimum wheelhead). However, when using the fired clay wheelhead, although it could be used to form pots, it quickly developed a series of cracks, requiring it to be wetted and shored up with clay. As the author was not likely to be making pots continuously each day, it was thought best to use a wooden wheelhead instead as it was more durable. Interestingly, the fired clay wheelhead found by Verner (1992; pp. 55-9 1995, pp. 27, fig 27a, pl 25) at Abusir had similar problems with cracking. It possibly had been repaired in antiquity as it had several holes which had been interpreted by the excavators as being repaired by threading a cord through it (Verner 1992; pp. 55-9 1995, pp. 27, fig 27a, pl 25).

However, there may be an alternative explanation for these "repair" holes (see Figure 6.14). Rather than being repair holes, these perforations could be connected to the use of the central hole, and deliberately added when the wheelhead was being made. Drilling holes after firing would damage a fired clay disk of only 10cm-3cms thickness. Instead, these could be emplacements for potters' bats,[42] which are commonly used by modern potters when throwing vessels to allow the finished vessel to be easily lifted off the wheelhead. The wooden dowels of differently sized bats could be placed inside the holes as desired, depending on the size of the vessel being thrown. The lower diameter (16.5cms) would fit most of the museum examples of mushroom-shaped stone pivots (see Chapter 2). Arnold (1993, p. 41) has suggested that bats may be being depicted in the tomb of Ty (see Chapter 3), providing useful secondary evidence. In Ty's tomb finished thrown pots on small disks are placed to dry on shelves above the potter throwing bowls on his wheel (Épron & Daumas, 1939, p. 71).

When putting the potter's wheel together, the socket is placed into the stand and the bolts fastened to prevent slippage. The socket and pivot working faces are lubricated with boiled linseed oil (a lubricant known to the ancient Egyptians and the most successful used during Powell's (1995) experiments). A circular slab of clay is rolled out and placed on top of the pivot, and the wheelhead is patted down onto it and coils of clay applied to the edges to secure it (see Figure 6.15). The author used standard earthenware terracotta[43] for all of the experiments in replicating miniature vessels.

The process of throwing on an ancient wheel is essentially the same as throwing on an electric one, although the motion

---

[42] A potter's bat is a flat surface, usually made of wood which the potter can place their pot upon when throwing. Often it is used as an aid to lift off larger pots and as a stand for drying. A centred, then flattened ball of clay is placed on to the potter's wheel and grooved with the fingertips when adding the bat to the wheel to make suction areas.
[43] CTM pottery supplies: Standard terracotta clay, blend of Etruria Marls, fires at 1080°C-1160°C to a light red colour as temperature is increased.

FIGURE 6.16: THE RECONSTRUCTED POTTER'S WHEEL. DRAWING: S. DOHERTY

FIGURE 6.15: ATTACHING THE WHEELHEAD TO THE CONCRETE WHEEL BEARING USING COILS OF CLAY. PHOTO: S. DOHERTY

force of the wheel is provided by the potter's left arm. These two bearings formed a thrust bearing to effectively absorb the force parallel to the axis of revolution. Placing a baked/fired clay or wooden wheelhead on top of the bearings added extra weight and increased the momentum of the spinning of the wheel. Pouring lubricant such as linseed oil (Powell 1995, pp. 316, 322, 331-334) in the socket prevented the tenon from locking inside the socket and maintained an even spin.

When using an electric wheel, the wheel is usually rotated at its fastest speed while the arms are locked into position and the clay is squeezed until no oscillation of the hands occurs and the clay is centred (see video at http://bit.do/potterswheel). When using the ancient wheel, the author (see video) tried to achieve similar results by positioning her right arm up against the leg so that it would stay straight and would only push against the clay. Notice however in Figure 6.17, that the seating position of the author is perhaps not the same as the representations of seated potters. This is probably due to lack of practice. The ancient potters are usually represented seated on a little bench and have their legs either side on the wheel or crouched just behind it, e.g. potter in the tomb of Ty (Épron and Daumas 1939) or the limestone statuette of Nikauinpu[44] (Breasted 1948, pp. 49, pl 45; Teeter 2003, pp. 21, 25).[45] When centring the clay, it is important to add plenty of water to reduce friction and make centring easier. However, the friction of the clay is initially quite strong as the author was only able to rotate the wheel 5 times before more water was needed. At the beginning of the experiments, the concrete wheel bearings had not been completely run in and one spin from the hand on the wheelhead only rotated the wheel 1½ times (approximately 15-20 rpm see Edwards and Jacobs (1986, 1987) above). In later experiments, particularly when using the granite wheel this rotation increased to 45 r.p.m.[46]

When beginning to open out the vessel it is possible to spin the wheel more times as plenty of water has been added. The rim of the vessel can be easily manipulated and changed throughout the throwing process and even during the finishing process when the vessel has been dried to the leather hard stage. Therefore, the existence of rilling marks on the rim of a pot may not be useful in representing a completely thrown vessel. Once centred, creating a pot on the concrete wheel bearings took about 11 minutes, but it is likely that Egyptian potters would have been far faster (see Figure 6.17). Similar hand-rotated wheels are still in use in Afghanistan and Pakistan as documented by Roux and Corbetta (1989), where potters are able to create pots in under 5 minutes. In contrast, coil pots can take much longer, depending on the finishing and drying times in between coil attachments.

The pot may then be trimmed of its excess clay, particularly at the base of the vessel as that is where clay often accumulates. Any likely tool can be used for this purpose, Blackman (1927, pp. 152, fig 180), in her ethnography, notes that the side of a thin square of iron perforated in the centre was used as a turning tool or rib by potters. Such makeshift tools have been found at the pottery in Lachish

---

[44] OIM 10628, Oriental Institute, University of Chicago.
[45] This position is still common amongst Egyptians today who are often seen squatted down resting on their ankles.
[46] R.p.m. or Rotations per minute was measured by applying a fixed point to the wheelhead and filming the wheelhead spinning. The film was then slowed down post-production. The amount of times that the wheel made a complete rotation was then counted, and an average per minute was taken.

FIGURE 6.17: (LEFT) THE AUTHOR HAS FINISHED CENTRING THE LUMP OF CLAY ON THE RECONSTRUCTED ANCIENT WHEEL AND IS COMMENCING OPENING OUT THE VESSEL WITH THE FINGERTIPS. (RIGHT) THE AUTHOR IS SHAPING THE BODY AND RIM OF THE VESSEL PRIOR TO ITS BEING REMOVED FROM THE POTTER'S WHEEL. PHOTOS: S. DOHERTY

FIGURE 6.18: THE POTTERY TOOLS FOUND IN THE POTTER'S WORKSHOP AT LACHISH. AFTER: TUFFNELL 1958, PL 215

including pebbles and shells for burnishing, an animal rib for trimming and smoothed pottery sherds for shaping (Tuffnell 1958, pp. 291-293, pl 215; Magrill and Middleton 1997, pp. 68-73; see Figure 6.18). Similar examples have been uncovered at the Amarna site Q48.4 (Rose P. J., 1989, pp. 89, fig 4.5). The author used similar items made of wood. Once the vessel was deemed sufficiently complete, the thrown pot was removed from the hump using twisted wire or string, either while the wheel was stationary or when spun.

Once made, the pot is left to dry for approx 1-2 weeks (in the UK, less in hotter climates) until it is "leatherhard" and ready to be fired in an electric kiln. The leatherhard stage is the last point during which the potter can alter the pot in some way e.g. adding handles, burnishing or finishing the rim (Nicholson 2009, p. 1) but in the case of the experimental wares, nothing was changed.

Experiment 2: Making a granite replica Potter's Wheel

Having tested the concrete prototype thoroughly and found that it could successfully be used to throw pots, it was decided to proceed one step further and make a granite version. This was constructed using the same template as the cement version (BM 32622) but using Mourne granite[47] (see Figure 6.20, Figure 6.21 and Appendix IV for plans). It was found that through continuous use, the cement version was gradually wearing down to the aggregate mix

---

[47] Made by S. M. McConnell's and Sons Ltd, Kilkeel, Co. Down, N. Ireland through the Cyril Fox Fund.

FIGURE 6.19: AUTHOR REAPPLYING LUBRICANT (BOILED LINSEED OIL PICTURED TO RIGHT) TO THE CONCRETE POTTER'S WHEEL REPLICA IN BETWEEN THROWING POTS. NOTE THE DARKENED WORKING FACES CAUSED BY INCREASED FRICTION TO THIS AREA. PHOTO: S. DOHERTY

FIGURE 6.20: THE CARVED AND HONED GRANITE REPLICA POTTER'S WHEEL BEARINGS. NOTE THE LUBRICATION DISCOLOURATION ALREADY STARTING TO FORM. PHOTO: S. DOHERTY

FIGURE 6.21: THE GRANITE WHEEL BEARINGS SET UP. PHOTO: S. DOHERTY

within the concrete body. This had resulted in an increase in friction, despite liberal application of lubricant.

The new granite wheel was much more effective once it was "run in" i.e. was rotated several times to ensure that it would spin freely with minimum oscillations. Presumably, the same would have occurred for the ancient potter, who when they first received a newly cut set of potter's wheel bearings would have had to spend a lot of time smoothing and rotating it before it would have been useful. Many of the examples in the museum collections are very smooth, shiny and have continuous rilling marks scratched into the working faces, presumably through continuous use; perhaps they were used through several generations. The granite wheel bearings significantly decreased the amount of time that it took to throw a vessel compared to the concrete version (5 minutes (45 r.p.m compared to 11 using the concrete version 30 r.pm.)) and allowed for more accurate results when compared with the museum examples.

Observations

Using the concrete and granite wheel bearings, the author was able to form several vessels in a similar manner to the miniature vessels. However, only through direct comparison with the museum examples could it be proved that the author had indeed manufactured the vessels to the Egyptian standard. One replica vessel was selected at random to compare to the archaeological. First, when considering the body of the two vessels, it was noticeable that both displayed similar features (see Figure 6.22).

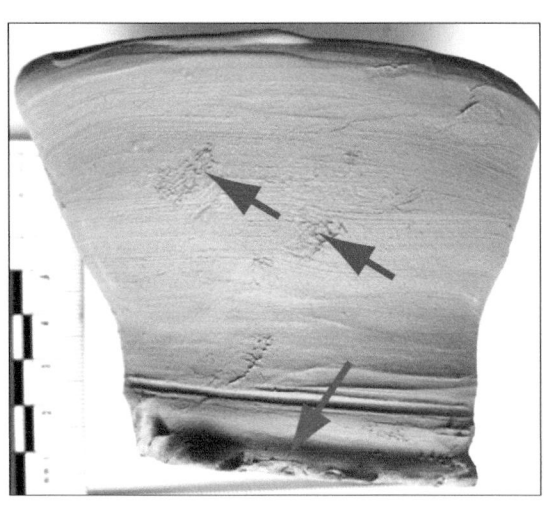 

Figure 6.22: (Left) The outside of the replica pot. (Right) the outside of the archaeological miniature vessel AN1895.766. Ashmolean Museum, Oxford. The upper arrows indicate sticky fingerprints or marks left when the vessel was lifted off the potter's wheel; the lower arrows indicate the uneven base as the pot was cut from the lump of clay attached to the wheel. Photos: S. Doherty

Both pots have concentric striations beginning at the base of the pot and continuing to the top of the rim. Each have remains of the fingerprints left behind when the potter was lifting the finished vessel from the potter's wheel. The stickiness of the clay that was used is indicative that throwing was the primary method of manufacture, as when throwing, potters tend to intermittently add water to the clay to ensure that the clay does not dry out and so that the potter's wheel can run smoothly. When coiling, it is better to refrain from adding too much water or else the vessel walls might collapse. The two pots also display evidence of being cut off the lump of clay on the potter's wheel as the bases are uneven. They are rather lopsided, evidence that they were not correctly centred on the potter's wheel.

Second, when one considers the inside of the vessels, one can again detect similarities (see Figure 6.23). At the inside base of the vessel, each have a hollowed out depression c. 0.5-1cm in diameter, otherwise described as a torsion crack or dimple, indicated by the red arrows in Figure 6.23. This was created when the potter first placed their fingers into the clay (see Figure 6.17) as centrifugal force is being induced to open out the vessel. The inner base of thrown vessels can sometimes exhibit another manufacturing mark known as s-shaped cracks, created if the potter later smoothes over the initial opening out depression with their finger. The clay later cracks as a result of excess water, which often pools at the base of the vessel during forming and is left there by the potter after the pot has been set aside to dry. This crack was not seen on the examples selected.

Another key feature, also reflected on the outside of the vessel, is continuous striations from the base of the vessel to the edge of the rim of the pot, indicated by the blue arrows in Figure 6.23. This is a crucial point, because if the rim of the vessels only presented these striations, then one would not consider this pot to be thrown, but merely rotated as centrifugal force is not induced.

Third, when one examines the bases of the vessels, again there are parallels between the archaeological and replica examples. Both exhibit string impressions, occurring when the base of the pot was sliced off using the string. However, on the replica miniature vessel, the base displayed straight lines or drag marks indicated by the white arrows whereas the archaeological example showed spiral string marks indicated by the yellow arrows. Upon further experimentation, it became apparent that the archaeological example was in fact cut off from the lump of clay while the potter's wheel was still in motion (see Figure 6.24).

Experiment 3: Using the Wheel bearings to Finish coil Pottery (V-rim bowl)

A final experiment was designed to independently replicate the V-rimed bowls that Courty and Roux (1995; Roux 1990; 2008; Roux & Courty 1997) created using the pierced potter's wheel bearings to shape rather than throwing vessels. They postulated that Near Eastern potters made coiled "roughout" vessels and then smoothed and finished

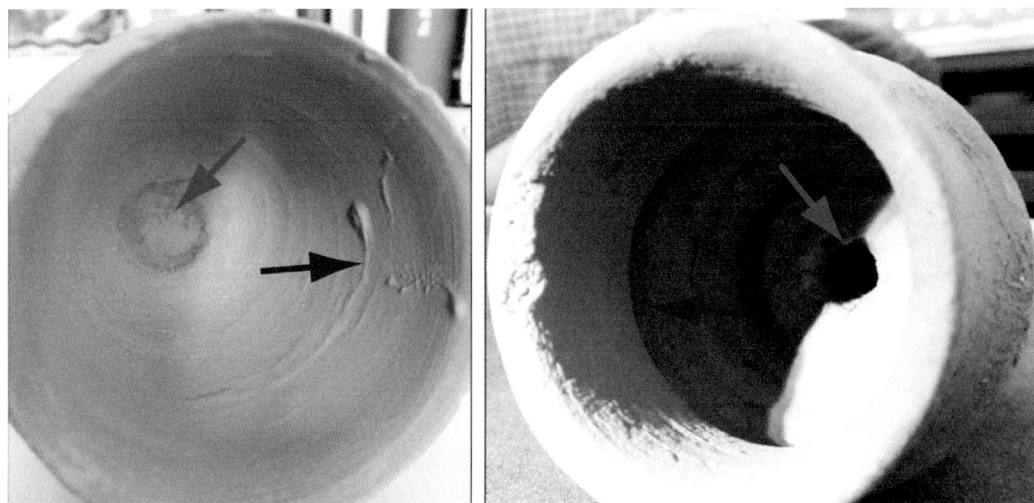

FIGURE 6.23: (LEFT) THE INSIDE OF THE REPLICA POT. (RIGHT) THE INSIDE OF THE ARCHAEOLOGICAL MINIATURE VESSEL. AN1895.766. ASHMOLEAN MUSEUM, OXFORD. THE ARROWS IN THE CENTRE INDICATE THE TORSION CRACK OR DIMPLE, CREATED WHEN THE POTTER FIRST PLACED THEIR FINGERS INTO THE CLAY TO OPEN OUT THE VESSEL; THE ARROWS ON THE EDGE OF THE POT INDICATE THE CONTINUOUS STRIATIONS CREATED BY THE SPINNING OF THE POTTER'S WHEEL. PHOTOS: S. DOHERTY

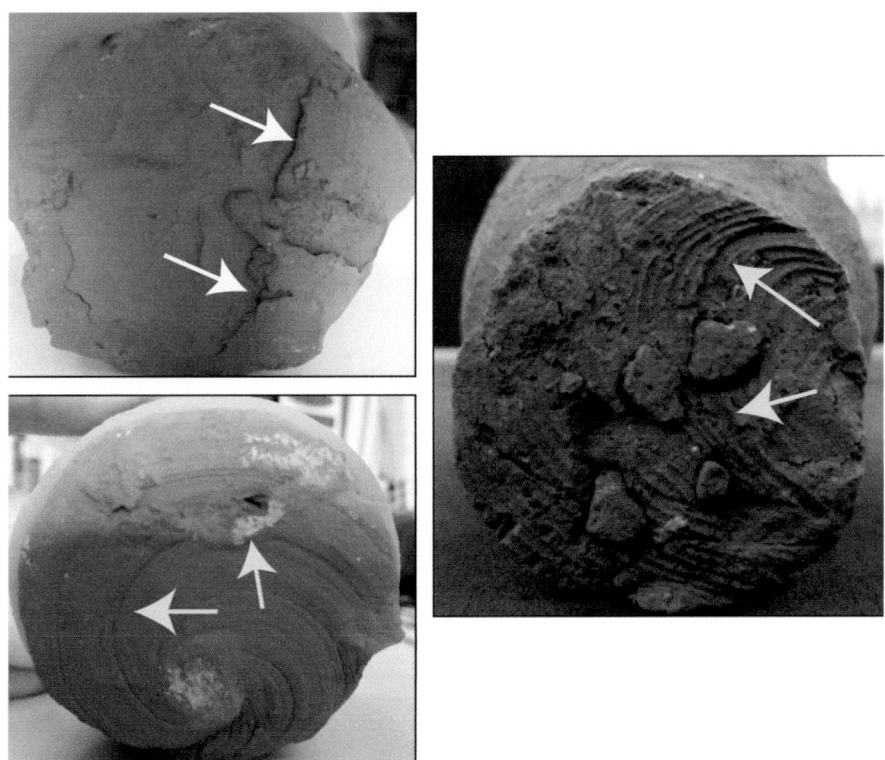

FIGURE 6.24: (LEFT) THE BASES OF THE REPLICA POTS. (RIGHT) THE BASES OF THE ARCHAEOLOGICAL MINIATURE VESSEL. AN1895.766. ASHMOLEAN MUSEUM, OXFORD. THE ARROWS ON THE UPPER REPLICA VESSEL INDICATE THE DRAG LINES CREATED WHEN THE POT WAS REMOVED FROM THE POTTER'S WHEEL WHILE IT WAS STATIONARY; THE ARROWS IN LOWER REPLICA VESSEL INDICATE THE SPIRAL LINES CREATED WHEN THE POT WAS REMOVED FROM THE POTTER'S WHEEL WHILE IT WAS STILL IN MOTION. PHOTOS: S. DOHERTY

the pots on a wheel (see Figure 6.25). This represented a logical step between using a support or turntable to draw up the sides of a vessel and the development of wheel throwing (Roux and Courty 1998, p. 748). A similar experiment was undertaken by Pelta with the pierced wheel bearing discovered at Tell Dalit (Pelta, 1996, pp. 171-185, see Chapter 2; Table 2.1 for details and Figure 2.7 and Figure 2.8). In these experiments with pierced wheel bearings, the speeds achieved (over 50 r.p.m) bearings were deemed not suitable for throwing (Roux and de Miroschedjii 2009,

  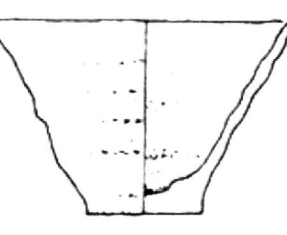

FIGURE 6.25: EXAMPLES OF V-SHAPED BOWLS, MADE BY ARRANGING COILS OF CLAY AND THEN THINNED AND SHAPED ON THE POTTER'S WHEEL. LEFT AND CENTRE: BM 125942; 1937, 1211.224 FROM TELL BRAK ©TRUSTEES OF THE BRITISH MUSEUM MIDDLE PHOTO: S. DOHERTY. RIGHT: PROFILE OF AN EXAMPLE OF THE V-RIM BOWL AFTER: ADAMS & NISSEN 1972, PG 309, FIG 6.G

p. 165). Other authors have suggested that 50 r.p.m. is sufficient, (Jacobs & Borowski, 1993, pp. 53-55), but most cite Rye's (1981, p. 74) 80-100 r.p.m. as a more suitable speed for throwing. With the granite bearings, the author was able to achieve these speeds and induce centrifugal force even at 40 r.p.m.

For V-shaped bowls, the primary method for shaping the pot was through coil, pinch or slab, and the potter's wheel is used as an aide to finish and thin the vessel so that the potter can stay in one place rather than have to move around the pot while forming it. The potter's wheel in the Near East at this period was not rotated sufficiently fast enough for centrifugal force to be achieved. This is similar to the "banding wheel" or "whirlers" used by modern potters. The method for creating the V-rim bowl was outlined by Courty and Roux (1997) as follows: (1) Vessel is built up with coils upon the potter's wheel. (2) The wheel is spun and the coiled pot is thinned and shaped. (3) The pot is cut off the wheel and the base removed. (4) It is placed on a mat to dry. (5) Finishing touches are added and the pot is smoothed.

It was decided to replicate the method concluded by Courty and Roux (1997) for "finishing V-rim vessels." Unfortunately, it was not possible to make a set of pierced wheel bearings. Instead, the pivot and socket granite wheel was employed in the use of a *tournette* or turntable and was not rotated more than 50 r.p.m. as outlined in the V-rim bowl experimental method designed by Roux and Courty (1997 see Chapter 2). Ten vessels were constructed using the above method; one was polished and burnished to compare finishing methods (see Figure 6.27). The observations are detailed below.

Observations

In many ways, the V-rim vessels created closely resembled the coil-built macroscopic details as outlined in Table 6.2. Internally, they contained a flat base, the edges of the vessel were smoothed and thinned through the spinning of the wheel (see Figure 6.26). Many of the coils were covered over by the clay during the oscillation of the wheel. This resulted in an even appearance as most of the traces of scraping and fingertip impressions had been removed (see Figure 6.26). This demonstrates a phase of pottery manufacture in between coiling and throwing. In Figure 6.27, the example shown was burnished and polished using a pebble after it was thinned and shaped on the potter's wheel, giving it a shiny and smoother appearance (see Figure 6.27). This also would have an effect upon the pores within the clay, sealing the vessel and making it impermeable to water if the vessel was used to store liquids in.

Summary

It is evident from the literature that Early Dynastic and Old Kingdom pottery is still frequently labelled as "wheel-finished" or "wheel rim rotated," when usually this is not the case and in fact the vessel is question has often been hand-built using coils. Through experimentation by making wheel thrown and hand-made (coil) pots, it

FIGURE 6.26: INTERNAL VIEW OF REPLICATED V-RIM VESSEL (UNFIRED). NOTE FLAT BASE AND SMOOTHED SIDES. IN THIS EXAMPLE, THE FINAL SET OF COILS ARE STILL DISTINGUISHABLE. PHOTO: S. DOHERTY

FIGURE 6.27: THE SMOOTHED OUTER EDGE OF THE REPLICATED V-RIM VESSEL (UNFIRED). THIS EXAMPLE HAS BEEN BURNISHED WITH A PEBBLE RESULTING IN ITS SHINY APPEARANCE. THE BASE OF THE VESSEL IS FLAT AND THE TRACES OF THE COILS HAVE BEEN LARGELY OBLITERATED.
PHOTO: S. DOHERTY

became evident that there were many differences between the techniques used for manufacturing the vessels, these were highlighted by using x-rays, and adding iron oxide spangles to the clay. Some coil pots did contain striations and spiral marks reminiscent of wheel-throwing, especially on the rim of the vessel. This implied that some form of rotation was occurring during the finishing of the coil vessel. However, it did not mean that the entire body of the pot was thrown, which is a very different process. When building coil vessels, the potter usually does not dampen the clay except occasionally when attaching the joins of the coils and when finishing the vessel. As such, the coil-made pot rarely exhibited sticky or wet finger marks. The profile of the pot often undulated with the coils rather than having a slick tapering profile as do the wheel thrown examples. Instead, the pots are placed on a support/bat which is rotated as the potter smoothes their fingers along the rim to even it out.

Having identified suitable early wheel thrown pots from the Old Kingdom in the pottery collections of the Ashmolean Museum, Oxford Petrie Museum of Egyptian Archaeology, UCL and Egyptian Museum in Cairo, they were then examined for any manufacturing marks. Such marks, (e.g. the spirals, scrapes and striations) left by the ancient potters were compared to the marks created when using the replica Egyptian potter's wheel. These marks appeared to match with the pots dating to the 4th Dynasty that were being fashioned on the wheel, confirming Bárta's (1995) original thoughts that miniature vessels were wheel thrown, apart from the underside of the base, and are likely to be the first pots to be thrown on a potter's wheel, crucially having centrifugal force being induced in order for throwing to be taking place. This has led to an increase in understanding and knowledge about the methods used to create such pots in ancient times. In later experiments, it was discovered that if the wheel was continuously rotated and a piece of string was applied to the base of the pot to cut it off from the hump, similar marks could be achieved on the base as noted on the archaeological examples. From these experiments, a list of manufacturing marks left on wheelthrown pottery has been created (see Figure 6.10) which will be useful for future identification of wheel-thrown pottery.

These experiments have advanced our knowledge by resolving the terminological problem of what constitutes a vessel thrown on a hand-spun potter's wheel when compared with a vessel which has been formed by coiling. Vessels which exhibit evidence of partial rotation on the rims, such as wavy handled jars or other storage jars, have not been thrown on a potter's wheel, but instead are likely to have been placed on an unmovable block or support such as a mat and then the pot rotated by the potter (as noted in the depiction of the pottery workshop tomb of Ty (Steindorff 1913, pp. 83-84; Épron and Daumas 1939, pp. 12, pl 71). Issues relating to the speed that a potter's wheel needs to achieve before it can be considered a "fast" versus a "slow" wheel can now be disputed. The author can find no such distinction, as the replica potter's wheel was successfully able to create thrown pottery at speeds lower than the suggested 50-150 r.p.m. (Rye 1981, p. 74), thus inducing centrifugal force, even at the speed of 20 rpm, not considered by Edwards and Edwards (1986, pp. 49-56) to be throwing. It is suggested that such terms as *fast* and *slow* wheel needs to be readdressed, if they should exist as a distinction at all. Near Eastern V-rim bowls were also recreated, through building up coils of clay, and then shaping and thinning the resulting vessel by slowly rotating the replica potter's wheel. This reproduced the method outlined by Courty and Roux (1997) and demonstrated a phase of pottery manufacture in between coiling and throwing. This was perhaps the first use of the potter's wheel in the Near East, before it was discovered that it could be utilised for true wheel-throwing inducing centrifugal force.

The practical use of the potter's wheel has been considered in this chapter, when it occurred and how the Egyptians may have used their wheels to throw pottery. Chapter 5 postulated that the origins of the potter's wheel was in the Near East, which then was used briefly in Buto to finish Canaanite colonists' coiled pots (in a similar manner to the V-rim bowl), seemingly to depart again when the Canaanites did. It was only with the reign of Pharaoh Sneferu in the Old Kingdom, that the potter's wheel began to be in use by indigenous Egyptians, but only for royal contexts. It has been established that miniature vessels were the first wheel thrown vessels. Using the replica prototype concrete wheel, the author was able to create a pot in 11 minutes using the granite bearings the author was able to make a similar pot in under 5 minutes. In Chapter 7 the throwing of miniature vessels on the potter's wheel will again be considered, but this time in relation to their development and spread within the court and to the wider communities within the Egyptian state. In the next chapter the author will consider the spread of the potter's wheel as its use became more widely accepted.

Chapter 7:

The Spread of the Potter's Wheel from Royal to Domestic Contexts

In this section, the instigation of the use of the potter's wheel and its links to funerary cults will be examined through analysis of the purposes for which miniature and model vessels were used. These vessels were used in chapels and pyramid temples to hold the daily offering. Chapter 6 was concerned with their manufacture and put forward the proposition that in Egypt they were the first vessels thrown on a potter's wheel. This chapter will consider the function and use of these miniature and model vessels. The first use of wheel thrown pottery seems to have been as vessels for daily ritual offerings in funerary cults dedicated to the *ka* (Gahlin, 2001) of a Pharaoh or private individual. These mortuary cults became important during the time of monumental pyramid building by the Pharaohs of the 4th-5th dynasties (c.2600-2300 B.C.). As part of their funerary pyramid complexes, these Pharaohs (beginning with Sneferu (c.2640-2604 B.C.)), built new temples specifically dedicated to the nourishment of their *ka* in the afterlife. These rituals required daily offerings of food, drink and other items to be presented to a representation of the *ka* of the deceased king by a priest. Private individuals also had chapels and needed to provide for their own *ka's* offerings. As these offerings occurred daily, a regular supply of small vessels produced in large numbers were required in which to place the food and drink, or to give symbolic offerings to the *ka*. Bourriau (1981, pp. 20, fig 11) suggests that these miniature vessels were entirely votive and never intended to be used as a container, apart from symbolically for the funerary cult.

Miniature and Model vessels

Miniature vessels were only used in specific funerary and religious temple complexes and their use in daily votive offerings was probably regulated by the elite. It is postulated that this was perhaps the reason why the potter's wheel (see Wheel terms in Figure 2.1 in Chapter 2) was brought in for an inherently Egyptian purpose. Rather than using the potter's wheel to create Jordan Valley style wheel-finished, coil, V-shaped bowls for elite ritual and settlement contexts (Roux, 2003; Roux & Courty, 2005), the Egyptians used this technology to create entirely wheel thrown miniature vessels. Akin to the Jordan Valley V-shaped bowls, these thrown vessels were made specifically for the increasingly important funerary cult that was controlled and managed by the elite administrative classes. The Egyptians were effectively borrowing the technological idea of the potter's wheel, but utilising it in an innovative way; that of throwing rather than shaping pottery. By the time the great pyramids of Giza were being built, the potter's wheel was used to create vessels for an additional purpose, but still within the funerary sphere. Evidence from excavations at Heit el Gurob, the village where the conscripted pyramid construction workers were living, suggests that the potter's wheel was utilised to supply the workers with eating vessels such as bowls and plates and these are known as CD7 bowls (Wodzińska, 2006). The use of the potter's wheel therefore still remained within the sphere of funerary elite administrative bureaucracy, but with an additional range of functions. Miniature vessels have a key role in determining how the use of the potter's wheel developed in Egypt and their origins will now be scrutinised.

Funerary Offering Rituals

When a deceased person was deposited within their tomb following their death and funeral, the tomb was not merely sealed and left. Relatives of the deceased, in particular the eldest son, were expected to make provisions for the continued well being of the deceased's *ka* through supply of regular offerings of food and drink to replenish those given at burial. Most tombs comprised of two areas, a subterranean burial chamber surrounded by storage rooms for offerings, and over the grave a brick mound or chapel (sometimes two) in which pairs of stone stelae were placed giving the deceased's name and titles. These mounds or chapels represented the primordial mound from which the creator god emerged and via which the dead king or person would be reborn and consequently they were the centre of the funerary cult where offerings were deposited (Taylor, 2001, p. 141).

By the 2nd dynasty, tombs contained evidence that complete meals were being offered, including pudding, cheese, and wine courses, the food laid out on plates, and cups for the wine. Cuts of beef, particularly the foreleg and head that were sometimes mummified, bread and honey cakes were popular items deposited in the tombs (Emery, 1962). These rites eventually extended to all members of Egyptian society who could afford to do so, but probably originated in the Pre and Early Dynastic tombs of the kings, as many contain large storerooms filled with provisions. Many of the tombs of the early kings at Abydos have small, open chapels with offering niches and stelae,[48] in front of which the offering rituals were re-enacted (Spencer, 1991, pp. 49-54) e.g. the tomb of King Djet as described by Kemp (1966, p. 13). By the 3rd dynasty, the offering chapels were incorporated within the mastaba[49] with a central recess for the statue(s) of the deceased person, known as the *serdab,*

---

[48] In later times the stelae took the form of an imitation door carved in stone, consequently they are often referred to as "false doors" which were believed to function as a real door in the afterlife so the spirit of the deceased, or *ba* could leave the tomb during daylight hours.
[49] From the Arabic word 'mastaba' meaning 'bench', for the massive rectangular structures found above many tombs in Saqqara, Giza and other places.

inside which the *ka* could live. In some cases these chapels became a complex series of rooms e.g. Saqqara mastaba 3518 with decorated reliefs and paintings of offerings (Seidlmayer, 1990; Reisner, 1934, p. 581). The statues were treated rather as the king must have been in life. As the pyramid texts record, every day the statues were ceremoniously woken, washed, and purified, the rituals of the opening of the mouth, eyes, and ears were performed and breakfast served. Then the statue was anointed with cosmetics and clothed, and finally a banquet lunch was provided (Spencer, 1991, p. 55).

There is evidence that these rites were formalised under the Old Kingdom Pharaohs. Instead of occurring within the mastaba, the offering chapels were moved to the pyramid mortuary temple, and the nourishment of the deceased was taken care of through magic and ritual using miniature vessels rather than by providing real-sized foodstuffs, as it would be impossible to provide fresh food offerings in perpetuity (Baines & Lacovara, 2002, p. 15).

The Step Pyramid of Djoser contained "dummy" structures designed specifically for the provision of offerings to the king's *ka*. These included a court of the *serdab* to the north, with statues of the king dedicated to the nourishment of the *ka* (see Figure 7.1) and a false mastaba tomb to the south, known as the "South Tomb." As the Pharaohs were said to have two or more *Kas* being part god, part human during life, perhaps these funerary structures were providing for

FIGURE 7.1: THE STATUE OF DJOSER'S *KA* FROM HIS SERDAB AT SAQQARA. EGYPTIAN MUSEUM, CAIRO.
PHOTO: S. DOHERTY

both (Spencer, 1991, p. 58) and may be a nod to the earlier "subsidiary" graves at Abydos not meant for the interred body, but rather as the symbolic resting place of the *ka*. This may also be the origins of the satellite pyramids rather than for the burial of Queens' e.g. the Satellite Pyramid of Khafre may have been used for the burial of the statue of his *ka* (Lehner, 2008, p. 126).

The next major stage in pyramid construction occurred during the reign of Pharaoh Sneferu. Alongside the building of a "true" pyramid without steps throughout several phases of building at Meidum, the pyramid temple was positioned to the eastern side of the pyramid, in accordance with the development of the cult of the sun god Ra and the rising sun. It retained the trappings of earlier offering chapels, as it contained an inner courtyard with two offering stelae and an offering table. There are no inscriptions, which is unusual, as without writing the name of the deceased, the Egyptians believed that they would not endure. This has led Lehner (2008, pp. 97-100) to suggest that the pyramid at Meidum was left unfinished when Sneferu died (c.2550 B.C.) as he constructed potentially three other pyramids, with the one at Meidum meant to be a cenotaph rather than a tomb.

From the beginning of the 4th dynasty, the outward appearance of mortuary ritual and the provision of offerings became more important than the provision of actual consumable offerings. Many tombs began to contain increasing numbers of miniature pottery vessels and dummy or model stone vessels, full-scale vessels devoid of any food and false doors. Symbolic offerings seemed to be the rule of the day, as there became increasing understanding that the funerary cult could not be easily maintained by succeeding generations, even those of royal status (Baines & Lacovara, 2002, pp. 15-16; Kemp, 2006, pp. 141-9; Shirai, 2005). Therefore, the offerings became symbolic, with wheel-manufactured miniature pottery, Meidum vessels and dummy model stone jars used to serve the cults of the deceased elites, which once offered, would magically transform into provisions for the afterlife. These vessels will be the next topic for discussion.

The Offering Triad: Funerary Model Objects, Meidum bowls and Miniature Vessels

At the same time as miniature vessels were being made on the wheel, a wide variety of model objects were also being created for use in funerary contexts. The use of model objects as part of the funerary furniture of Dynastic Tombs was common in Ancient Egypt, particularly during the Old and Middle Kingdoms, although they had their origins in Predynastic (Swain, 1995, pp. 35-7). Model vessels and objects were thought to become functional items in the afterlife that could be used by the deceased to equip themselves with the necessities of life. A wide variety of models were made such as boats, cattle, servants (see Chapter 3), houses, craftspeople, workshops and soldiers, but models of votive pots, bowls and other containers were also popular items of funerary goods. These were made of wood, pottery, faience or stone, often calcite or travertine, and the goods occurred in tombs across the Egyptian elite classes, both royal and private individuals (Breasted, 1948; Tooley, 1995, pp. 8-11).

Representations of model objects occurred in texts e.g. *htp di nsw* offering formulae, as pictures of model vessels on tomb offering scenes and stelae, and in physical material objects such as pottery, stone and metal vessels. The use of model objects rather than life-sized made both economic and space-saving sense. It would be less expensive to provide a wooden or pottery version of something such as a herd of cattle which would have required feeding until death when needed to be replaced, or else later be expensively mummified. By being represented in a smaller form, it was believed that the model contained the essence of the real life object(s) and would later be useful to the deceased in the afterlife (Swain, 1995, p. 35). However, Bourriau (1981, p. 117) suggests that the Egyptians had no need to economise their funerary goods as such, as by magical means the model could become the item represented so clearly in the model. Models are usually found only in elite contexts, and are often very elaborately painted in imitation of glass etc. These objects are often varnished to preserve the pigments and were probably very expensive to buy, so became part of the standard funerary repertoire.

There were two differing kinds of small vessels used by the Egyptians for funerary cult rites; model and miniature vessels, found usually as bowls or plates made of pottery or stone, and one large serving vessel, the pottery Meidum bowl (see Figure 7.2). The model and miniature vessels were seldom taller than 7.5cms or wider than 6.5cms, and usually occurred in funerary and ritual contexts alongside full sized vessels, especially Meidum bowls (see Figure 7.2 and Figure 7.3). Both miniature vessels and model stone vessels were known from the Predynastic Period, but miniature vessels initially occurred as wheel-made pottery during the reign of Pharaoh Sneferu, first king of the fourth dynasty (c.2640-2604 B.C.) (Bárta, 1995, p. 15). They were mass-produced and designed to be used once and then discarded, as many thousands have been excavated in pristine condition (some 10,000 at Dahshur (Fakhry, 1961, p. 135)).

The Egyptians probably thought of miniature and model vessels as two distinct types, although they were related in shape and function (Allen, 2006, p. 20). Miniature, model and full sized Meidum vessels all occurred together in tombs and other funerary architecture, in both private and royal contexts e.g. Queen Heterpheres's tomb, wife of Sneferu and mother of Khufu (Reisner & Stevenson Smith, 1955, pp. 76-7, fig 100-1 miniature, 143-4 model, 99 full scale). It is likely that the Egyptians regarded each vessel type as having a distinct purpose, and that the model and miniature vessels were not merely cheap replacements of the full sized examples (Allen, 2006, p. 22). Model vessels may copy existing full sized vessel forms or they may be in shapes restricted to the model vessel corpus, usually in a non-functional capacity. Miniature vessels can be compared to vessels that are full sized, but which have been reduced in scale whilst retaining their functional ability, and probably could contain a token amount of liquid or grain (Faltings, 1989, pp. 153, note 43).

Unlike most other pottery types, Meidum vessels often have an S-shaped profile (see Chapter 6, Figure 6.4) and carinated rims usually with a red slip of ochre or hematite applied prior to firing and then burnished. They are regularly depicted in tomb and temple scenes associated with eating, food presentation, and feasting, and are often placed on pottery stands covered with a lid made of basketry (Wodzińska, 2006, p. 411). They occur mostly in

FIGURE 7.2: LEFT: A DUMMY STONE MODEL VASE MADE OF CALCITE. 6TH DYNASTY. NOTE IT IS COMPLETELY SOLID. UC69832. MIDDLE: A WHEEL THROWN MINIATURE POTTERY VASE MADE OF NB2 CLAY (VIENNA SYSTEM). 4TH DYNASTY, MEIDUM. UC17609. RIGHT: MEIDUM VESSEL RED SLIPPED POTTERY 17.5CM IN DIAMETER, 4TH DYNASTY, MEIDUM. UC17636. UCL PETRIE MUSEUM OF EGYPTIAN ARCHAEOLOGY

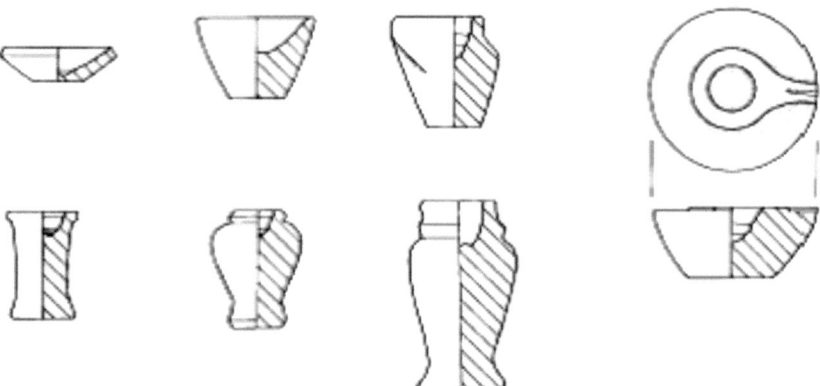

FIGURE 7.3: MODEL VESSELS MADE OF CALCITE FROM GIZA TOMB G 7440 Z, 4TH DYNASTY HEIGHT 1.7-6.4CM, DIA 2.1-6.3CM. HARVARD UNIVERSITY OF FINE ARTS EXPEDITION, 1927 (27.1483-1591) D'AURIA, LACOVARA, ROEHRIG (1988, P. 77, FIG 37-43) AFTER: (ALLEN, 2006, PP. 20, FIG 1)

funerary contexts from the 2nd to the 6th dynasty but some have been excavated from Old Kingdom settlement sites (Op de Beeck, 2004, p. 239). These three pottery types will be described and discussed in further detail below.

Model Vessels

Model vessels have no functional ability, many are almost solid (see Figure 7.2 (left) and Figure 7.3) and are sometimes referred to in excavation reports as "Scheinbeigaben," mock dishes or dummies (Martin-Pardey, 1984). Their solid nature would indicate that the outward form must be the most significant thing about the vessel, and the content was probably implied by their shape e.g., *hs* vases or *nmst* jars (D'Auria, Lacovara, & Roehrig, 1988, pp. 77-8).

A model vessel was usually made of stone, although some occur in copper such as the examples found by Reisner in the tomb of Heterpheres (1913, pp. 62, fig 16). Models were designed to be smaller versions of larger forms that were used for votive offerings, they often were not hollow, or only had a relatively shallow depth and therefore would not have a functional ability, but be important symbolically, in a similar manner to the hieroglyphs of the offering formulae *htp di nsw* (see Figure 7.3). These formulae occurred from the Old Kingdom onwards and were inscribed upon doorjambs, furniture and coffins. From the reign of Sneferu, these formulae began to be incorporated into the tombs of private individuals (Bárta, 2005, p. 182). They were prayers asking the king to make a representative offering to either the god Osiris or Seth on behalf of the deceased, and then for offerings of bread, beer and every good thing required by the deceased's *ka* for their happiness in the afterlife (Gahlin, 2001, pp. 166-7). This sometimes-exhaustive list often included alabaster vessels, while the model jars perhaps represented the 3D form of the formulae (e.g. see Figure 7.2). Offerings in the form of model food made of faience or cartonnage were thought to be magically equivalent to real food, and more likely to be preserved.

Model vessels have been recorded in contexts dating to the Merimde Beni Salame culture in Lower Egypt (4800-4200 B.C.) e.g. at Ma'adi (Rizkana & Seeher, 1988, p. 67). As they were not made of pottery, model vessels were not made using the potter's wheel, but were made of stone using drills such as the Twist Reverse Twist Drill. Most early model stone vessels were made of calcite,[50] or travertine as these are relatively soft stones (Aston, 1994, p. 42). Working harder stones such as basalt requires the use of a cold smelted copper tipped drill which was not invented until Naqada II (c.3600-3200 B.C.) (Amer, 1933; Stocks, 2003, pp. 11-2).

Miniature Vessels

Miniature vessels were similar in size to model vessels being up to c7.5cms tall and up to c6.5cms wide. Miniature pottery vessels of the 4th dynasty were produced on a potter's wheel and unlike model vessels had an interior volume, and could probably contain a token amount of liquid or grain and therefore retained their functional capability (see Figure 7.4). However, they may have not been used functionally, but as a symbolic item (Bourriau, 1981, p. 117). They can be compared in shape to full-size vessels and possibly were emulating them. The fabric of the miniature vessels are almost exclusively Nile B2 in the Vienna system (Nile silt clay with black core/narrow red core, inclusions of straw, sand mica and limestone see Appendix II). They were mass-produced, but were destined for a specific cultic and funerary sphere, and were used daily as part of offering rituals to the dead and then discarded after one use (in a similar manner to the Mesopotamian bevelled rim bread bowls (Beale, 1978; Goulder, 2010)). During the 4th dynasty, these miniature vessels seem to only occur in elite and royal ritual contexts, such as pyramid mortuary temples, tombs and chapels, beginning with the foundation deposit at the pyramid of Sneferu at Meidum (Clayton, 1994, p. 45; Dodson, 1995,

---

[50] Also known as Egyptian Alabaster.

p. 27). These vessels were created to serve and nourish the *ka* of deceased royal and private individuals with a token offering of food and drink. Once the rituals were finished for the day, the used vessels were then deposited into large rubbish pits such as those at the pyramid temple of Menkaure found by Reisner (1931), perhaps later ending up being incorporated into wall linings or foundation deposits when the pits were cleared to make way for more (Charvát, 1981).

Miniature vessels have frequently been overlooked by excavators as they often occur in great quantities and can be thought of as quite crudely fashioned, perhaps owing to the speed by which they were made (see Figure 7.4). At Meidum, the types of these miniature vessels most commonly found were saucers with a small foot, which were wheel-made with the surface wet smoothed and very pale brown (10 YR/7/3). The fabric type was described as (Aiib) coarse Nile clay, low fired with a black core and light brown surface zones, and with inclusions of a lot of long chaff pieces and some sand, which could correspond with Nile B1 or B2 in the Vienna system (see Appendix II)

To the Egyptians, aesthetically pleasing pottery may not have been the priority criterion for manufacture; the most significant thing was that the pottery achieved what it was supposed to. Since these miniature vessels were likely to be single use only for the purpose of rituals, the potters who made them were probably not overly concerned with producing appealing pottery. However, it seems that these small pots may be highly significant in terms of the uptake of the new technologies of the potter's wheel, updraught kiln and the beginnings of administrative control of the pottery industry in Egypt. By being able to control how pottery was made, what it looked like and what it was used for, the elites were able to manage their craftsmen in new ways. The potter's wheel was probably permanently located within the confines of a temple workshop area, and in a location where the output of the working potter could be scrutinised by the overseer (see Chapter 2).

In relation to pottery, mass-production rather than standardisation of form seems to have been the significant thing. Scribes usually noted the amount/weight of grains used to produce beer or bread rather than the amount or

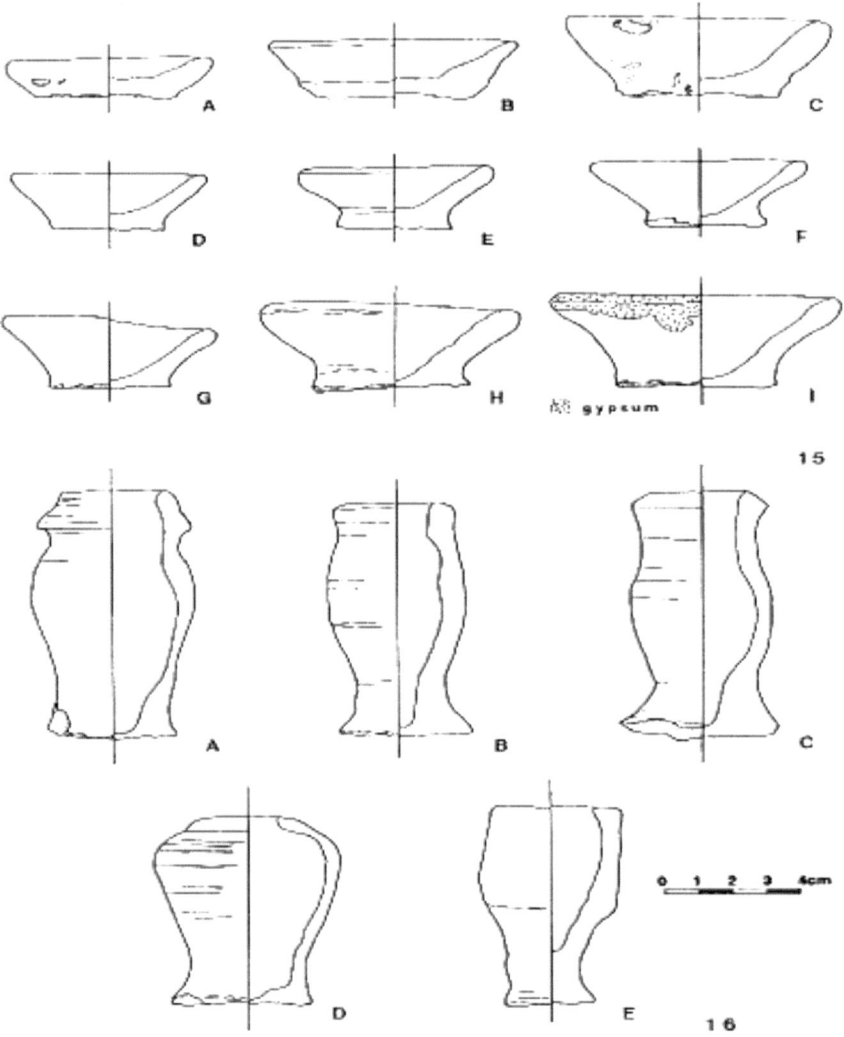

FIGURE 7.4: 4TH DYNASTY MINIATURE VESSELS FROM MEIDUM
AFTER: EL-KHOULI, PG 44, PL 50, POTTERY TYPES 15-16

size of vessels used to contain them "scribes and potters... were worlds apart," noted Kemp (2006, p. 174). Scribes seemed to be unconcerned about pottery quality control, in terms of vessels shape standardisation, but focused rather on the speed of production. Warden (2010, pp. ix, 127-128), in her examination of beer jars and bread mould volumes of Old Kingdom date, suggested that there was no standardisation of such hand-made vessel forms so therefore there was no all encompassing state-run system for the pottery industry. She surmised that Pharaoh was likely to have been an employer for those who worked at his state-run institutions such as the mortuary temples and pyramid towns which were funded by taxation. Private individuals could then function as employers in the other non-state sponsored economic areas (Warden, 2010, p. 128). If the potter's wheel began its use-life through state sponsorship, it would therefore be expected that wheel thrown vessels would be solely found within elite state-organised contexts. From the 3rd-4th dynasty, the first wheel thrown pottery, occurs only in the most illustrious of state-run projects, that of pyramid mortuary temples and mastaba chapels in the form of miniature vessels. This suggests that potters specialising in the use of the potter's wheel were for at least one dynasty kept within the confines of the Pharaoh's control in state-controlled temple workshops and later this specialism was disseminated[51] to private individuals running their own estates.

The resulting use, rather than the impetus of the potter's wheel had the advantage of increased speed of production when compared with the production of hand-made pots, particularly if the pots were thrown off the hump as 5th dynasty depictions suggest (see Figure 7.8). In addition, the use of the elite hard stone such as basalt and granite for the potter's wheel bearings to fabricate pottery miniature vessels may also be significant. The use of basalt as an elite state quarried product will be considered further. There is some evidence that Predynastic hand-made miniature vessels were directly emulating basalt vessels of the same period, indicating the close early links between basalt and pottery production which will be discussed below.

Miniature Vessels and their links to Basalt

More so than any other Egyptian pottery, miniature vessels seem to have close links to basalt and stone vessel production. Both types of vessels have similar shape and style characteristics, occur in similar contexts in the same sites and dates, and may have had related functions. At first glance, the production of vessels made of stone and those made of pottery was likely to be separate, but with the addition of the potter's wheel, this may not have been the case, as the wheel bearings for the wheel were made of stone and therefore a stonemason was required to make them. The likely close links between stonemasons and potters utilising the potter's wheel to make miniature vessels will be considered in this section.

Miniature vessels are usually made of pottery and occur in funerary and mortuary contexts, although some examples occur in stone (e.g. UC15611 and see Figure 7.5). The first miniature vessels have been noted in Petrie's *Corpus of Prehistoric Pottery* as black polished and fancy wares (1921, pp. 6, pl XIX nos 80a-82) dating to early Naqada II. Miniature vessels of this period were almost exclusively fired to a black colour when excavated at Ma'adi and other Predynastic sites, indicating their firing in reduced conditions. Many of the miniature vessels have a raised base, some with lug handles and are fired black. Lug handles are generally unknown on the red polished and reddish brown large sized vessels of this period and it has been suggested that these early miniature vessels were emulations of basalt jars, which served as a cheaper alternative or "toy" version of the larger vessels, or containers of rare and precious materials and therefore were associated with cultic offerings, usually restricted to cemetery sites[52] (see Figure 7.5) (Rizkana & Seeher, 1987, pp. 45-6).

Miniature jars may have originally been used as a container of cosmetics, as an example was found containing a red greasy substance at Ma'adi (Amer, 1933, p. 29); others have been noted placed on top of bigger vessels as lids (Junker, 1912, p. 29). Reisner (1931, p. 216) noted in the 4th dynasty mortuary temple of Menkaure that black polished pottery was still part of the pottery corpus, although it was rare compared to red wares. He postulated that the vessels were a cheaper alternative to stone vessels and considered that the black polished pottery was an impractical ware due to its relative softness (Reisner, 1931, pp. 216, part 9 fig 66). Although miniature vessels were known and used for cultic practices from the Predynastic Period, it was not until the 4th dynasty that they were made on any grand scale. The close early links to basalt emulation may be the reason why the Egyptians later chose to create miniature pottery vessels on a basalt wheel bearing potter's wheel. It may be that close early links to the Levant and Mesopotamia were the source of this technological inspiration.

There is some debate regarding the extent of Mesopotamian influences upon the Egyptians during the Predynastic and Early Dynastic periods (Frankfort, 1941; Moorey, 1987) particularly in relation to the production of stone vases as discussed in Chapters 4 and 5. These stone vases are thought by some to be of Palestinian/Canaanite origins; perhaps they were brought to the port town of Buto[53] by merchants and then spread to Ma'adi and the delta (Faltings, 1998a, pp. 30-2). There is also evidence of colonisation of Palestine by Egyptians (Porat, 1992, p. 435). It seems likely that given the difficulty of procuring,

---

[51] Possibly during the 5th dynasty when the Meidum bowl begins to be fashioned on the potter's wheel and when depictions of the wheel first begin to appear in private individuals' tombs, e.g. Ty (Épron & Daumas, 1939).

[52] Although this may be due to lack of excavation of Predynastic settlement sites.

[53] Buto is likely to have been a port town during the Predynastic period; changes in the Rosetta branch of the Nile over time have led to the changes in the coastline.

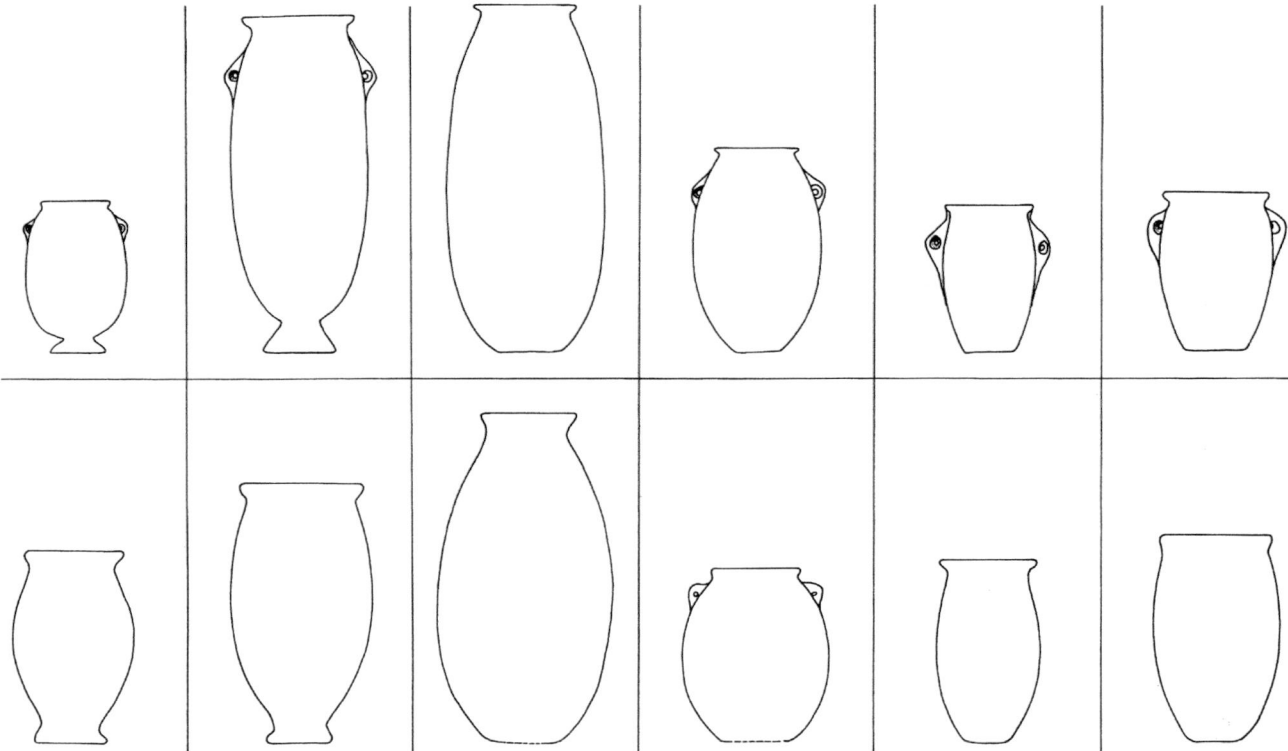

FIGURE 7.5: SHAPE COMPARISON OF PREDYNASTIC (NAQADA I-II) BASALT STONE VESSELS (TOP ROW) AND MINIATURE POTTERY VESSELS (BOTTOM ROW) FROM THE SITE OF MA'ADI. AFTER RIZKANA & SEEHER (1988, PG 68, FIG 16)

working and finishing basalt in contrast to pottery, that the easier alternative was for pottery to emulate the basalt vessels.

A second debate concerns whether pottery emulated stone or vice versa, which is still uncertain. The style and shape of basalt jars change from a lug handled barrel shape (such as Figure 7.5) to that with a ring base. These changes occur in both Lower and Upper Egypt in Late Naqada I; prior to this, they were solely a Lower Egyptian trend (Rizkana & Seeher, 1988). The changes indicate the beginnings of standardisation in both stone and pottery vases prior to the unification after Naqada III c.3100 B.C., with cultural characteristics' blending and homogenising as the north and south have more continuous contact with each other. The use of basalt for stone vessels, like the miniature vessels, appears to be purely destined for the funerary or religious sphere. These vessels were crafted to be only placed in the tomb; they do not seem to have been used domestically and rarely have any use marks. This might indicate that they may never have been used by the deceased and were made solely for the funerary market, to be deposited next to the corpse upon burial. It may also be the reason why the stone vase maker's workshop was located next to the temple of Horus at Hierakonpolis, namely, to provide stone vases for pilgrims making offerings to the god (Kemp, 2006, p. 196) (see map of Hierakonpolis in Chapter 4, Figure 4.10). Another workshop dating to the Old Kingdom has recently been located near Sheikh Said. (Willems, et al., 2009); perhaps the temple personnel were regulating the vessel production. This conveniently seems to coincide with the first use of smelted and cast copper in the Predynastic town of Ma'adi, located south of the apex of the Delta (Amer, 1933; 1936). Stone vessels also seem to have been restricted to elite burial contexts.

The use of stone vessels may have initially been driven by the elite seeking an alternative to pottery for use in their tombs. Stone vessels gradually became increasingly ornate in design and it seems at least in part that the potters were encouraged by the fashion and design of the stone vase makers to make pottery versions. Elite status may be conferred upon a person through means of personal achievement or inheritance. Inherited status is generally thought to be a sign of a more complex society, as it procures benefit for a limited number of elites and represents a significant stage in socio-economic development (Bard, 1988, p. 52; Wilkinson, 2001, p. 29).

Pyramid Building and Miniature Vessels

During the 4th dynasty, there were major changes in the religious and administrative structure of the Egyptian society, which almost certainly had an impact on pottery production. It seems that pyramid building may have been intrinsically linked to the increased complexity of the Egyptian state. The reign of Pharaoh Sneferu in particular was significant. Although not many images of Sneferu survive, in terms of monumental pyramid building and religious focus, he was a pioneer much earlier than Akhenaton's religious "revolution" (1351-1334 B.C.). Sneferu changed the orientation of religion towards the sun god Ra with the new title *s3 Rˁ* "son of Ra." He built true pyramids without steps, and aligned his pyramid complexes (Meidum and Bent and Red pyramid of Dahshur) in a linear sequence along an east-west axis following the path of the sun (Hannig, 2003). At the western end, he located his burial place within a pyramid and in the east; he located his mortuary temple for the nourishment of his *ka* in the form of everyday offering rituals, served in miniature vessels. The Pharaohs no longer used their mortuary temples as eternal re-enactment of rituals of mortal kingship, but instead became demi-gods as sons of Ra, and when they died became one of the gods and enacted the eternal cycle of the sun's renewal (Robins, 1997, p. 45) (although they still considered themselves as the representation of Horus on earth to later become Osiris in the afterlife). These developments implied a change in the relationship of the king to the sun and Ra, with the true pyramid shape represented as a beam of sunlight, relating to and aligning with the *ben-ben* stone at Heliopolis (Edwards, 1986; Jeffreys, 1998; 2010, pp. 112-3; Kemp, 2006, pp. 85-8). By changing the focus of the Egyptian belief system, Sneferu was also changing the political, social, and economic system and using his pyramid to impress his people (Lehner, 2010, p. 85).

It had previously been thought (Aldred, 1980, p. 58) that Pharaoh Huni, the predecessor of Sneferu, was the builder of the pyramid at Meidum; however, the consensus currently among scholars is that if Sneferu did not start the building work at Meidum, he most likely completed it (Clayton, 1994, p. 45; Dodson, 1995, p. 27). However, Lehner (2008, p. 198) considers that it is unlikely that Sneferu completed a pyramid of Huni's, as usually a new Pharaoh would not finish their predecessor's tomb. Since the name of Huni has so far not been located at Meidum, whereas Sneferu's name is found in the surrounding mastabas and in later New Kingdom graffiti (Petrie, 1892, pp. 40, plates XXXIII & XIX), it is more likely that the pyramid of Meidum was built by Sneferu. Writing in the *Academy*, (1891, p. 376; 1892, p. 9), Petrie mentions the base of a serpentine statuette being found inscribed with the town name "*tˁt-sneferu*" dedicated by a woman named "*šneferu-khˁti*," which would attest to the area being built by Sneferu.

Sneferu and the Mass Production of Miniature Vessels

During the reign of Sneferu, several changes in tomb architecture occurred for both private individuals and members of the royal family. This possibly mirrored changes in the economisation of the funerary industry due to a state economy, which was overburdened because of pyramid construction (Krejčí, 2000, pp. 467-84). Before the reign of Sneferu tombs had been large; up to several thousand metres squared e.g. Mastaba of Neferma'at and Atet at Meidum was 8160 m² (Harpur, 2001). However, at the cemeteries of Dashur, the size of tombs decreased to a standard size of 600 m². The substructure of private tombs also decreased to a single underground chamber, perhaps indicating the economic strain that Sneferu's persistent pyramid building created upon his elites as he redirected resources towards their construction. For the first time, the offering formulae *htp di nsw* occurred in non-royal tombs, the written metaphysical version of the physical offering version of the miniature vessel (Bárta, 2005, p. 182). The more obvious demonstrations of wealth noted by the size and decorations in the mastabas of Meidum at Dahshur became less discernable with elites allowed only the same size and similarly decorated mastabas.

The uses of miniature vessels first discerned under and within the funerary structures at Meidum, were probably part of the elite economisation process. The use of a new technology often coincides with some form of social and/or economic pressure (van der Leeuw, 2002, pp. 239-240), which Sneferu created through his construction of at least three pyramids and his dissatisfaction with their designs. This would have put the Egyptian economy under considerable strain, and it seems that alongside the economisation of the size and decoration of mastabas, there was also an economisation in the funerary cults and the vessels used during offering rituals. Although Sneferu allowed his elites greater freedom of expression of their lives and their status at Meidum e.g. large Mastaba tombs (Goedicke, 1979, p. 121), he gradually forced his elites to display their wealth in less conspicuous ways. This was probably driven by the need for resources to be redistributed to ensure the construction of his tombs.'

Part of this process was the beginning of the mass-manufacture of miniature bowls, plates and so-called Meidum bowls, which were created to imitate the more expensive stone vessels. Op de Beeck (2004, p. 243) believes that the Meidum vessel may have its origins in the 1st dynasty. Bárta (2005, p. 182) suggests that Meidum vessels were designed exclusively to emulate stone vessels for the funerary cults of wealthy officials and held liquid or viscous offerings, possibly milk or dairy products (Op de Beeck, 2004, p. 23). These miniature vessels were designed for the upkeep of the daily cults of votive offerings, with miniature plates for food and bowls for drink. It is also possible that the miniature vessels continued the Predynastic tradition of emulating stone vessels. However, the pottery miniatures during the time of the Old Kingdom were fired in oxidising conditions, using NB 2 clay and being fired to a red colour. Consequently, they were probably no longer emulating black basalt, which may have been the case during the Predynastic as most miniature vessels of that time were fired black in reducing conditions (first noted by Petrie (1921, p. XIX)).

The increased use of miniature vessels are perhaps indicative of a rising administrative class which augmented the demand for their own tombs and associated funerary cults. From the 5th dynasty onwards there was evidence of the king's funerary cult continuing for several decades after the king's death e.g. a cult of Sneferu is recorded until at least the reign of Pepi II c.300 years later (Shirai, 2005,

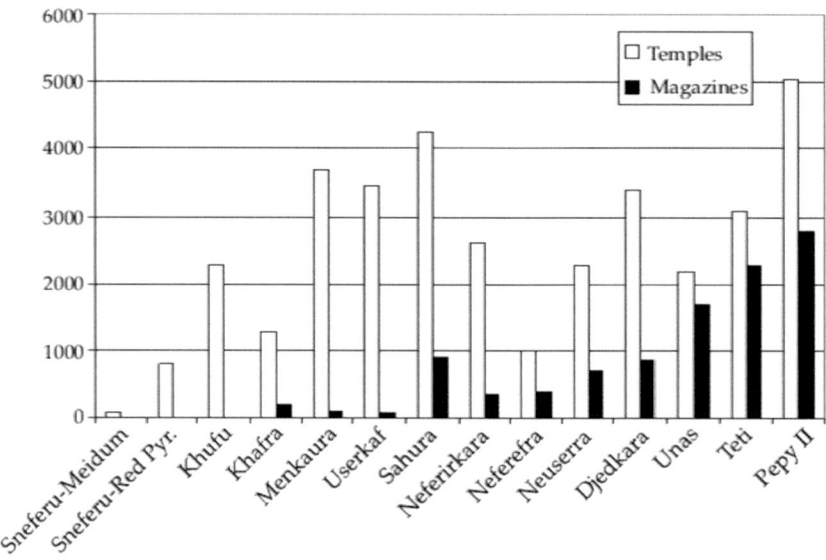

FIGURE 7.6: BUILT AREA OF THE MORTUARY TEMPLE OF OLD KINGDOM KINGS COMPARED TO THE AREA TAKEN UP BY STOREROOMS WITHIN THE TEMPLES. AFTER BÁRTA (2005, PG 184, FIG 4)

FIGURE 7.7: ARROWS INDICATE G. A. REISNER'S 1930S POTTERY SPOIL HEAP STILL VISIBLE TO THE SOUTH OF KHAFRE'S PYRAMID, CLOSE TO THE CAUSEWAY OF MENKAURE'S PYRAMID. INSET (LEFT) SOME OF THE MINIATURE VESSELS LOOSE ON THE SURFACE PHOTOS: S. DOHERTY

p. 151). This perhaps indicates that the pyramids should not be viewed as isolated entities, which were locked up and left once the Pharaoh was safely interred and ensconced amongst the gods for all eternity. Rather, they should be seen as large complexes perpetuating long after the "death" of the Pharaoh, continually celebrating his human life through daily offerings in his mortuary temple. The 4-5th dynasty intensification of the mortuary cults can be noted in the increased sizes of storerooms within the mortuary temples (see Figure 7.6), the increased size of the mortuary temple relative to the pyramid, and probably an increased number of priests specifically dedicated to the mortuary cults (Bárta, 2005, p. 184).

Pits within the temple complexes often contained thousands of miniature vessels indicating their use for daily offerings e.g. at Menkaure's mortuary temple at Giza many were found amongst the general debris of the chapels, offering rooms (III 2), 40-50 on floors of magazines (nos 16-18), some under the walls of the second temple and in the dump heaps thrown out from the chapels of the surrounding mastabas. Beside the entrance to the north of the mortuary temple, Reisner excavated a deposit of several thousand miniature vessels (1931, pp. 228-9, fig 79-80) and his pottery spoil heap is still visible today (see Figure 7.7).

Manufacturing Mass-Produced Ritual Vessels

Many early complex societies experienced a transition period in which highly decorated, labour intensive pottery was replaced by mass-produced forms of lower aesthetic appeal and less labour intensive pottery e.g. bevelled rim bowls in Mesopotamia, *bedjᶜ* bowls in Egypt as moulds for bread (Chazan & Lehner, 1990, pp. 21-35). It has sometimes been thought that these pottery decoration changes coincided with some form of cultural or social collapse. These changes possibly reflect changes in pottery use, and that pottery was not so important as a form of artistic expression (Trigger, 1983, p. 64). Changes in pottery may be an indicator of increasing cultural complexity as the king and administrative classes begin to control craft production (Johnston, 1987). In the case of Egypt, this seems to occur in very specific contexts: the use of *bedjᶜ* bread bowls made around a conical former and miniature vessels made on the potter's wheel (see Figure 7.8), both of which are represented by elites in their mastabas e.g. the tomb of Ty, where the production of pottery and bread are placed in the same scene, but differing registers (Steindorff, 1913). Miniature vessels and Meidum bowls, although mass-produced were designed for use in very specific contexts for cultic and funerary offerings and as such the elites were ensuring that their manufacture was also closely controlled. The manufacture of these vessels will now be discussed in detail.

Methods of Production: Miniature Vessels

Throwing consists of shaping a mass of clay on a quickly revolving horizontal disk. The mass is centred by the pressure of the hands, after which the clay is pulled up into a hollow form and shaped steadily by the pressure of the hands on both sides of the hollow. It requires practice and the correct manipulation of the clay with the hands to deal with the centrifugal action of the spinning wheel to prevent the pot from collapsing (Ruscoe, 1963, p. 18). Many of these processes can be reflected upon the finished pot. Often, archaeological ceramicists focus on the rim and the neck of a vessel, rather than the base or the body of the vessel to identify how the pot was made and to identify its type. However, these results can be misleading, as potters sometimes rotated the neck of a pot which was otherwise hand-made. The rim of the vessel is continuously manipulated throughout the shaping process and is usually the last thing to be finished.

Rye (1981, p. 74) has suggested that the appropriate speed for throwing is between 50-150 rpm, being inversely proportional to the diameter of the vessel at the point where pressure is applied. However, Birks (1979, pp. 13-15) suggests that the potter's wheel does not necessary have to be rotated quickly to achieve a thrown pot. One revolution per second is sufficient to place quite a lot of strain on the clay and to execute the necessary shaping of the vessel. The crucial point is that the wheel must revolve smoothly. During the 4th dynasty, Egyptian potters probably would not have been able to achieve high speeds on their hand-rotated wheels, yet they were creating miniature vessels that could only be described as thrown. The manufacturing marks upon the vessels as described above suggest that miniatures underwent all the processes involved in throwing and these marks are recognisable by modern potters today (Joan Doherty 2010 *Pers. Comm.*). The manufacturing stages are displayed in detail on later Middle Kingdom wooden models of potteries and in tomb scenes such as those at Beni Hasan (see Chapter 2) described by Newberry (1893; 1894). The clay was first prepared by an assistant through wedging with the feet and/or hands to remove air and extra water was added (see tomb models in Chapter 2). It was then made into cones and passed to the potter. The wheel was spun and the cone was dropped on to the wheel as close to the centre as possible. By already fashioning the clay into a cone,

 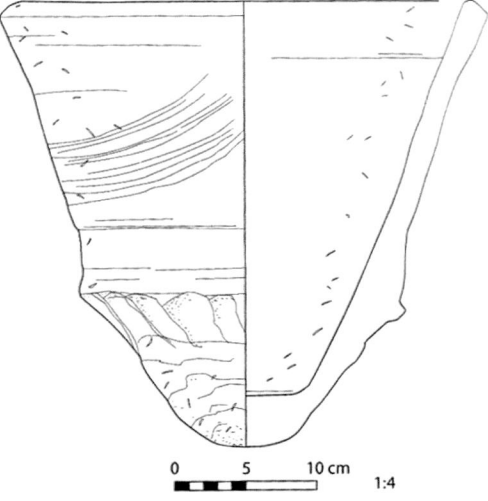

FIGURE 7.8: LARGE CONICAL *bedjᶜ* BREAD MOULD MANUFACTURED AROUND A CONICAL FORMER
WODZIŃSKA 2009C, LEFT COLOUR PL 6; RIGHT PG 142 FIG 67

the assistant was ensuring that the potter could commence throwing immediately and not have to knead the clay to remove air (see Figure 7.9). This would be an important time saver, especially if the potter was relying upon the strength of their left arm to rotate the wheel whilst shaping the vessel with the other hand (see video at http://bit.do/potterswheel).

FIGURE 7.10: THE RILLING MARKS ARE QUITE CLEARLY DISCERNIBLE IN THIS MEIDUM VESSEL SHERD FROM BUHEN, PARTICULARLY ALONG THE RIM. OLD KINGDOM 10CMS (L) x 14.5CMS (W) ©PETRIE MUSEUM OF EGYPTIAN ARCHAEOLOGY UC20101. PHOTO: S. DOHERTY

FIGURE 7.9: THE EXPERIMENTAL WHEEL SET UP WITH PRE-PREPARED CONES OF CLAY, READY TO BE USED FOR THROWING. PHOTO: S. DOHERTY

The potter would begin to open out the clay using a finger, and then gradually draw the thumb and finger closer together through pinching in one continuous movement until the vessel began to form. Once the vessel was opened out, the processes of shaping and collaring or narrowing of the diameter could begin, perhaps using a wooden tool, pebble, or shell to aid this process. Once finished, the vessel was cut off the hump of clay with string, possibly by holding one end of the piece of string taut and touching the other end of the base where the cut is desired. The revolving wheel carried the string around the vessel and the other end is pulled causing the string to cut through the vessel freeing it, and leaving the spiral pattern (see Chapter 6).

Methods of Production: Meidum Bowls

Meidum bowls may have initially been made and shaped on a hump or former rather like the bread *bedjᶜ* bowls or Mesopotamian bread bevelled rim bowl (Arnold, 1993, pp. 21-2; Chazan & Lehner, 1990). Meidum ware bowls of the 4-5th dynasties often display a clear distinction between manufacturing marks on the rim and those on the body of the vessel. The rims often display fine parallel lines that are caused by rotating the vessel (see Figure 7.10 and Figure 7.12).

The inside surfaces of the vessel's body is smooth, while the outside displays traces of scraping, indicating that the body was probably shaped on a former or hump and pressed down until the required shape was made. This would explain the thinness of the body of the vessel and the overlaps in clay noted by Vandiver and Lacovara in their xeroradiographic studies (Vandiver & Lacovara, 1985, pp. 80, fig 18a-b). Arnold (1976, pp. 17, pl. 4a-b) considers that it is only by the 6th dynasty that Meidum bowls were made entirely on the potter's wheel (see Figure 7.11).

Yet the potter working on a wheel in the 5th dynasty tomb of Ty at Saqqara seems to be fashioning a Meidum bowl with a carinated neck (Steindorff, 1913). Similarly, the limestone statuette of a potter, possibly from the 5th dynasty tomb of Nikauinpu at Giza [E10628] Oriental Museum, Chicago (see Chapter 3) also appears to be making a Meidum-style bowl on his wheel (Breasted, 1948, pp. 49, pl 45). However, it could simply be that the artists making such representations were simply depicting a standard "bowl" rather than recreating accurately a specific type of bowl.

Wodzińska seems to have found evidence that possibly during the reign of Menkaure, a variant of Meidum bowls, known as CD7 and a type that are perhaps unique to Giza,[54] may have undergone the transition from handmade to being thrown on the potter's wheel (see Figure 7.13). These bowls were produced on a vast scale in one location (the village of the 4th dynasty pyramid tomb builders) in a very short space of time and only during the time of the fourth dynasty (2600-2450 B.C.). CD7 bowls are a variant of Meidum bowls made of fine and medium fine Nile silt clay covered with a white wash (see Figure 7.13 and Figure 7.14). This is an unusual feature as Meidum bowls usually have a red slip applied before firing. These CD7 bowls seem to have had a very specific purpose for feeding the workforce of the 4th Dynasty Pharaohs' pyramid builders. Wodzińska has also found evidence that these CD7 bowls are unique amongst the pottery assemblage at Giza in that they were perhaps initially hand made and then later made

---

[54] Although similar vessels to the CD7 bowl have been found at Sheikh Said.

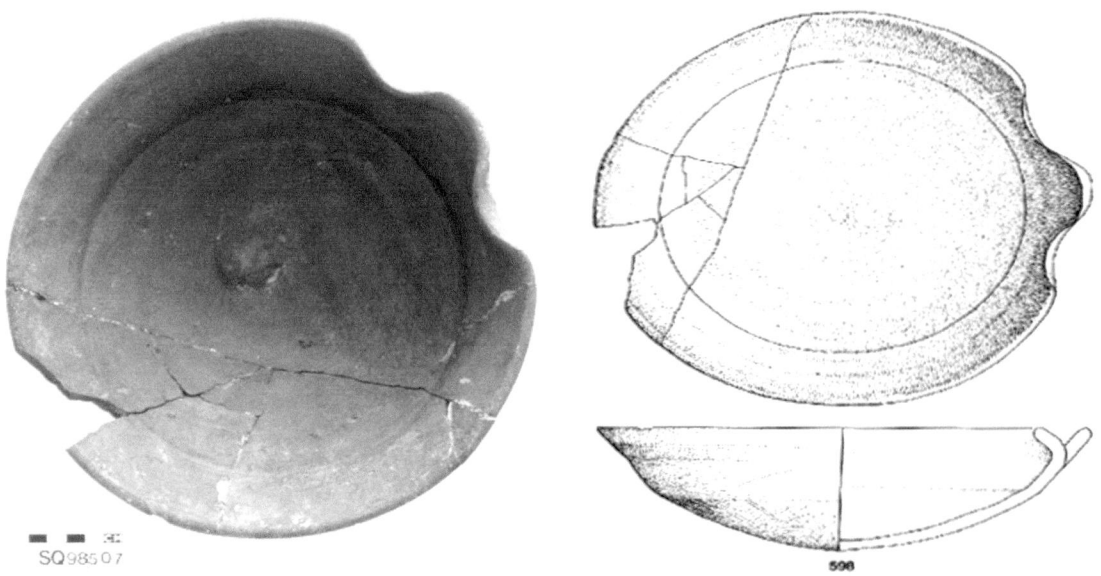

FIGURE 7.11: 6TH DYNASTY BOWL WITH SPOUTED RIM, FROM SAQQARA SQ98-507 TYPE 598. CLEAR SIGNS OF BEING THROWN ON THE POTTER'S WHEEL (SPIRAL AT THE BASE) RIM DIA 29.5-24CM, HEIGHT 7.7CM, NILE B1. AFTER RZEUSKA (2006A, PG 276, PL 117 AND HTTP://BIT.DO/POTTERSWHEEL)

FIGURE 7.12: CLOSE UP DETAIL OF A MEIDUM BOWL RIM SHERD SHOWING THE RILLING MARKS SIMILAR TO FIGURE. OLD KINGDOM, BUHEN. 5.1CMS (L) X 9.5CM (W) ©PETRIE MUSEUM OF EGYPTIAN ARCHAEOLOGY UC20091. PHOTO: S. DOHERTY

on the wheel. Many of these bowls apparently show clear signs of being rotated on the wheel as there are concentric striation marks on many of the rims and shoulders of the vessel, and the base is often irregular and trimmed (Wodzińska, 2006, pp. 405-429, see Figure 7.14 left).

However subsequent inspection of digital close-up photographs of the vessels by the author and Wodzińska suggest that this may not be the case. The CD7 vessels do not seem to exhibit the spiral striations, base cutting marks or the other characteristic marks associated with wheel throwing identified in Chapter 6. A brush was used to white wash the vessels after firing, and the brush strokes were applied in sweeping circular fashion using a 3-4cm brush, which at first inspection makes the vessels appear thrown. Wodzińska identified that many of the rims and the shoulders of the vessels exhibit striation

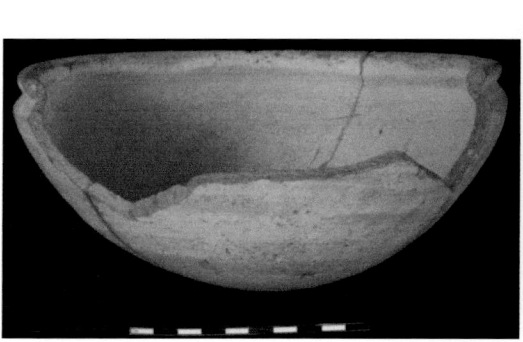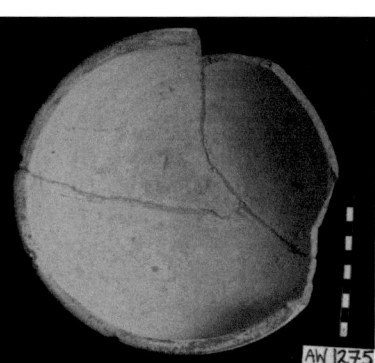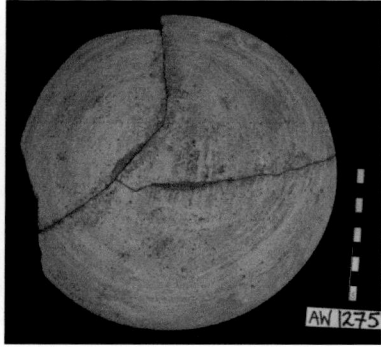

FIGURE 7.13: THREE VIEWS OF THE SAME CD7 VESSEL AW1275, FROM HEIT EL GHUROB, GIZA, REPRODUCED WITH KIND PERMISSION OF ANNA WODZIŃSKA, GPMP, AERA. PHOTO: A. WODZIŃSKA

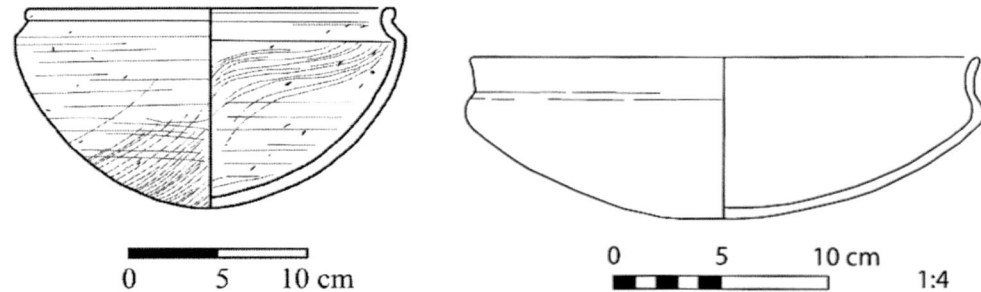

FIGURE 7.14: LEFT DRAWING OF CD7 BOWL MADE OF NILE CLAY, SCRAPED BASE AND COATED WITH WHITE WASH AFTER WODZIŃSKA 2009A, PG 238 FIG 18. RIGHT: EXAMPLE OF MEIDUM BOWL FROM GIZA, RED COATED AND POLISHED. CARINATED BOWL WITH ROUND SHOULDER AND ROUNDED BASE. AFTER WODZIŃSKA 2007, PG 301, FIG 11.2. BOTH EXAMPLES DATE TO 4TH DYNASTY (VEREECKEN, 2011, PP. 285, FIG 9)

marks, but no two examples are exactly the same (2006, pp. 405-429). However, as detailed above, this does not signify throwing, but just that the upper part of the vessel was smoothed and rotated, probably on a support, rather than the wheel. Unfortunately, it seems that the reign of Menkaure was not when the next stage of wheel production of Meidum bowls was instigated. If Do. Arnold (1993) is correct in her proposition that sometime in the 5th dynasty the potter's wheel begins to be used to create a greater variety of pottery types, beyond those used in the royal or funerary cult sphere, further analysis of provenanced Meidum bowls needs to be undertaken.

Due to the "luxury" nature of the Meidum bowl and their possible long use life through inheritance, lack of provenance in previous studies, and the automatic assumption by previous scholars that all pottery deposited remained in its primary position, it is often difficult to date these bowls (Op de Beeck, 2004; Rzeuska, 2006a, pp. 380-1). In many cases, the Meidum bowl appears to have been mostly deposited within the burial shaft e.g. shaft 1bis in the mastaba complex of Merefnebef (Rzeuska, 2006a, pp. 494-6, pl 188-9) which would suggest that they are possibly more likely than most funerary cultic vessels to be moved around or redeposited by visitors to the tomb.

Despite the introduction of the potter's wheel from the 4th dynasty for miniature vessels, other vessel forming techniques were not forgotten. Indeed some vessels such as beer jars continued to be made through coiling, and bread moulds by being moulded over a core or patrix (Wodzińska, 2009c, p. 147). In the 6th dynasty, many jars, bowls and plates of both open and closed forms are made on the wheel, with the exception of beer jars and bread moulds/trays. By the First Intermediate Period (7-10th dynasties c.2181-2025 B.C.), this hand-made manufacturing is completely overtaken by the potter's wheel, which is used to create almost all pottery types, with the exception of some of the coarser wares and some of the largest vessels.

The function of the pottery played a key role in determining how the pot would be made. It determined the type of clay used, whether Nile or Marl, and the shape of the vessel. The coarsest clays would have been selected for vessels such as bread moulds (Nile E see Appendix II and Figure 7.8) or cooking pots (Marl C1 see Appendix III). As these were often very thick walled c3-7cm because they required further heating during the food preparation process, it would have been easier for the Egyptian potter to make such vessels by hand. Pottery of the same shape could be made differently depending on the type of clay selected. Once a technique for forming a particular vessel is settled on, it would have taken quite a while before it would have changed, given that most vessels would have been made along a production line sequence as depicted in the various tombs scenes outlined in Chapter 2. This could explain the time lag between the inception of thrown pottery in the 4th dynasty for funerary and cultic small vessels, and its spread to larger and greater varieties of vessels during the 5-6th dynasties. The Egyptian potter needed time to learn their new craft, to train apprentices and to understand the strengths and limitations of the new machine. As the potter's wheel was made from basalt, it required a stonemason to procure the stone and work it for the potter, presumably at some expense. Consequently, it may have taken time before all potters were willing to take on this new technology, and perhaps required the sponsorship and support of their local wealthy noble or estate owner to purchase it for them. As ethnographic studies have shown, the potter is often amongst the most traditional craftworkers, once they work out the most effective way of undertaking a technique, they probably would see no reason to change. Craftspeople do not have many technological choices and are limited by the resources and machinery available and the traditions of craft production already developed e.g. South American mould made pottery makers use moulds made by specialist mould makers rather than making their own moulds (Pritchard & van der Leeuw, 1984, pp. 11-12; van der Leeuw, Papousek, & Coudart, 1991, p. 147).

Summary

It appears that the Egyptians borrowed the technological idea of the potter's wheel from the people of the Near East, but used it differently from previous usage. The Egyptians used the potter's wheel for first time to throw miniature vessels within the sphere of elite funerary administration. These miniature vessels have played a key role in determining how the use of the potter's wheel developed in Ancient Egypt. These miniature pots exhibit all of the six characteristic marks of throwing as defined in Chapter 6, and are therefore wheel thrown. The increased use of the miniature vessels from the beginning of the 4th dynasty during the reign of Sneferu (c.2640-2604 B.C.) onwards was for symbolic purposes, and was only sustainable for royal contexts. In this Chapter, the author has endeavoured to analyse why the potter's wheel came to be used during the 4th dynasty, in what form it was used and for what purposes. The contextual evidence of the vessels was assessed to establish how the potter's wheel was used to create pottery. It is likely that the potter's wheel was used to create vessels for the elites, and so it was found that, at least initially, wheel thrown vessels would only occur in elite contexts such as in ritual or funerary offerings. Since Chapter 6 examined the pottery of the early Old Kingdom to ascertain when the potter's wheel was in use and what pottery types potters were creating with their wheels, this Chapter has considered in what contexts they occurred. Early wheel-made vessels occurred in similar cultic and funerary contexts in Levant and Mesopotamia (Courty & Roux, 1995) and it appears that the Egyptians adopted this new technology to produce items in similar contexts (funerary and cultic). However, the Egyptians did so by a fundamentally different method, never before seen in the ancient world, namely by using the wheel to throw rather than finish vessels.

Chapter 8:

Conclusion

The aim of this present work has been to understand the reasons why the potter's wheel came to be invented and when and how it was used by ancient people and for what purpose. The potter's wheel is often thought to have originated in Mesopotamia in the 4th millennium B.C. and subsequently its use spread to the Levant and Egypt, but little analysis has been undertaken as to why this occurred, or how its use came to be so widespread (Freestone & Gaimster, 1997, p. 15; Kuhrt, 1995, p. 22; Pollock, 1999, p. 5; Simpson, 1997a, pp. 50-5). This research has sought to find the evidence on which this supposition is based through examining the archaeological and secondary sources (described in Chapters 2 and 3) and considering uncovered examples of pierced basalt wheel bearings from 4000 B.C. e.g. Tel Halif (Dessel, 2009, pp. 20-22, fig 7; Jacobs & Borowski, 1993) and potter's clay wheelheads dating to 3000 B.C. e.g. Ur, South Iraq (Simpson, 1997b, pp. 50, fig 1) to pivot and socket basalt and limestone examples in Egypt e.g. Amarna (Rose 1989). As outlined in Chapter 1, this present work has sought to answer the following (1) determining when the potter's wheel was introduced into Egypt, (2) establishing in what contexts wheelmade pottery occurs, and (3) considering the reasons why the Egyptians introduced the wheel when a well-established handmade pottery industry already existed.

In many ways, as argued in this book, the Ancient Egyptians had apparently no need of the potter's wheel, given they had already established highly successful pottery production techniques to produce vessels such as the Black topped wares or the Decorated "D-ware" pottery of the Predynastic Periods. These were hand-made, expertly fired in screen or box kilns e.g. at Hierakonpolis 11C (Baba, The Pottery Kilns at HK11C Revisited, 2006; Harlan, 1982) to create startling oxidised/reduced contrasting colours and decorated with red ochre slips or in the case of D-ware, paints (Petrie, 1921, pp. 1-8, 31-7). The use of the potter's wheel is almost automatically associated with the mass-production of pottery by archaeologists, because of the association that has been noted in ethnographic contexts, particularly when using "off the hump" throwing techniques. However, as Chapter 6 has shown, the potter's wheel even when being used to throw off the hump, can also be employed for a wide variety of finishing processes at slower speeds. As Roux and Courty (1998) have demonstrated, the use of the potter's wheel in the Levant was not for throwing, but for finishing coil made pots. In many parts of the world, particularly the Americas (Litto 1976), large amounts of pottery can be easily and quickly made using moulds. Rather than utilising the potter's wheel to mass-produce pottery, it appears that the Egyptians (as ethnographic studies have also suggested e.g. Nicholson 1995a) created a small range of vessels but using a greater skill to produce.

The adoption of the potter's wheel as a tool for rapid production of large numbers of vessels would have required significant changes to the entire pottery production process. Throwing on a wheel places constraints on the clays and tempers that can be used, as finer clay pastes need to be developed. Modifications to firing techniques and the development of the kiln would be instrumental in this process, allowing finer mixes of clay to be fired evenly. The areas of the world that have abundant fine alluvial clays e.g. the Levant, Mesopotamia and Egypt would therefore be more likely to invent the potter's wheel, once the economic and social conditions were in place to encourage specialists to concentrate on their craft, as has been demonstrated in this book.

With a variety of pottery making techniques literally at their fingertips, Chapter 4 investigated why the Egyptians decided to use the potter's wheel. The usual assumption is that the potter's wheel was instigated in order to create standardised mass-produced utilitarian wares (Bourriau, Nicholson, & Rose, 2000, p. 142; Blackman, Stein, & Vandiver, 1993, pp. 63-7; van der Leeuw, 2002, pp. 238-288). However, as this book has argued, this may not be the case in terms of the first usage of the potter's wheel, even if it was ultimately employed in mass production. The initiation of such a technology often requires some sort of impetus from another source such as the royal courts (Papazian, 2005, p. 75) or temples (Janssen, 1975, p. 183) before it can be instigated. The potter's wheel enabled its inventors to apply their knowledge of the shape of wheel (known from carts, lathes, waterwheels etc) to a new piece of machinery and specialisation of a craft that heretofore had been restricted to household or part time production. The potter's wheel also enabled the potter to make more symmetrical vessels in a greater variety of forms. Through its spinning, the potter was able to control the thinness of the walls of the pots to create a more even profile, which was its most important use for the creators of V-rimmed vessels in Levantine sites in the Negev region e.g. Abu Hamid, Beer-sheva, En Gedi, and Halif (Commenge-Pellerin, 1987; 1990; Dessel, 2009; Perrot & Ladiray, 1980; Ussishkin, 1980).

In Chapter 2, the current state of the literature relating to potter's wheels was assessed, and the problems that many scholars had with terminological misreadings were identified. Upon closer examination of the literature, it became apparent that scholars had yet to determine when the potter's wheel was invented. The literature reveals a variety of differing opinions regarding the date for the first use of the potter's wheel. The prevailing opinion seems to be that the invention of the potter's wheel could only have coincided with the beginnings of the Bronze Age, and the first use of working stone and copper/bronze tools. Potter's wheels were made from a range of different materials- baked clay, stones such as basalt or limestone- which would have required different tools to work and procure the bearings. Scholars also had problems describing the wheel and its many different forms and there was evidence of substantial confusion because of the differing terms to describe the same process that are currently used by modern potters and archaeologists studying ceramics. For the modern and traditional potter, for example, turning is not considered the same as throwing. Rather it is the scrapings, "turnings," or dry clay left after being removed from the vessel, usually from the base after it has dried. In archaeological literature, the term "wheel-throwing" has been used to refer to almost any type of rotational device in pottery making. Similarly, the potter's wheel has had problems in translation, with the terms *tournette* as "pottery disc" or wheel and *tour* or *tournage*, as "potter's wheel bearings" or "slow wheel" both being distinct terms in French. Such French-English classifications still occasionally occur within the archaeological literature and continue to cause confusion. Another term was applied to the hand-spun wheel as "slow" or "simple." Technically, throwing should only be used to refer to pivoted wheels rotating at speed for a considerable amount of time (Rice 1987, pp. 132-134) as described in Chapter 6. This author suggests that, to avoid this confusion between terms, the early potter's wheels that were used for hand throwing should simply be called the "hand-spun potter's wheels."

The outcomes from practical reconstructions of wheel bearings have been examined and how these different wheels have performed when pottery of differing types has been made on them. Provenanced potter's wheels (as detailed in Chapter 2, Table 2.1) have been analysed in term of material, dimensions, style and technical performance. In addition, the literature detailing the underlying manufacturing processes involved in throwing have been reviewed. It is suggested that the size of the diameter of the wheelhead will influence whether or not sufficient spin is achieved to enable throwing. Previous experiments making and throwing pottery using replica (and actual) excavated examples from Egypt and the Levant have been discussed. These experiments seem to indicate that Near Eastern Archaeologists consider the potter's wheel to not have been utilised for throwing, whereas professional potter working at Amarna Powell (1995) suggested that the potter's wheels excavated at Amarna would have been capable of throwing. When these experiments are analysed in detail, the speeds being achieved by the potters would have induced centrifugal force i.e. between 80-150 r.p.m (Rye 1981), and therefore could be considered throwing.

The secondary evidence for potter's wheels derived from images from tomb scenes, wooden models, and limestone statuettes was outlined in Chapter 3. From the authenticated tomb wall scenes dating to the 5th-6th dynasties, it is evident that potters were attached to estates of Egyptian royalty and nobility e.g. Ty, which pushes back the date of the first use of the potter's wheel in Egypt. There is a range of different types of potter's wheels depicted. Such scenes should be viewed with caution since they are often embedded with multiple, often-symbolic meanings and should not always be read as simply being representative of "everyday" activities (Kamrin, 1999; Walsem, 2005, p. 69). However, the evidence would suggest that some scenes do appear to represent accurate depictions of everyday life and could be used as a source of ethnographic information e.g. fishing and preparing fish (van Elsbergen, 1997; Nicholson and Doherty, 2014). The same could be said to be the case with specific pottery workshop scenes (Nicholson and Doherty, 2014). The statuettes and models dating from the 5th dynasty in particular are very similar to those depicted on the tomb walls. The wooden model workshops dating to the First Intermediate Period are very detailed and suggest the use of tools. The written manuscripts dating from the 4th-6th dynasties provide evidence for the first written evidence for the potter's wheel.

The potter's wheel is often associated with male as opposed to female potters, based upon ethnographic and historical data. This view seems to be confirmed in the tomb scenes described in this book. However, the association between male potters and the potter's wheel is probably less related to the technical skill of the male than to the social and economic conditions encouraging its use. Male potters typically dominate the craft in societies where large-scale production occurs without the use of the potter's wheel e.g. South America (Litto 1976). The organisation of production of pottery seems to be the key process involved for the gender shift in the use of the potter's wheel. The use of the potter's wheel allows for the increased specialisation of the entire pottery production process. With increased production, there is often a division of labour between the stages, with the most skilled potter using the wheel, apprentices processing the clay, wedging and aiding the potter at the wheel as noted in the Middle Kingdom tomb models in Chapter 3, Table 3.1. These processes can only realistically be undertaken in a workshop, with the potter's wheel permanently in place in the ground of the workshop, and a kiln outside to fire the vessels and aid vessel-firing survival rates.

Although a variety of experiments have been undertaken by different scholars on the practical use of the potter's wheel e.g. (Amiran & Shendov, 1966, pp. 85-87; 1984, pp. 107-122; Edwards & Jacobs, 1987; Hope, 1987; Pelta, 1996), none of them considered that the potter's wheel in the hand-spun form could be used to produce wheel thrown pottery, with the exception of Powell (1995). Authors have suggested that 80 r.p.m. is sufficient to throw pots (Amiran & Shenhav, 1984; Rye, 1981, p. 74). Edwards and Jacobs (1987, p. 52) achieved 15-20 r.p.m. whereas Powell (1995, pp. 309-335) proved that you could use a replica potter's wheel when she threw a variety of pots and bowls of New Kingdom types, achieving speeds of over 133 r.p.m. This author was able to independently confirm Powell's (1995) results in Chapter 6 since when using her replica potter's wheel she was able to achieve similar results.

When the potter's wheel was first utilised in the Near East, it was apparently only used for finishing and thinning coil-built pots. One of the most significant points addressed in Chapter 4, was whether or not centrifugal force is being induced when the potter's wheel is used to finish coil-built vessels. For V-shaped bowls, the primary method for shaping the pot was through coil, and the potter's wheel is used as an aide so that the potter could stay in one place rather than have to move around the pot while forming it as described in Chapter 6. In the Near East at this period (c.3500 B.C.), the potter's wheel was not rotated sufficiently fast enough for centrifugal force to be achieved, as the coils are clearly visible (Courty & Roux, 1995; Roux & Courty, 1997; Roux, 2003). Therefore, it could be argued that although the Mesopotamians invented the potter's wheel, they did not utilise it to achieve its full potential for throwing pots until much later and after the Egyptians had developed it to achieve this purpose. The invention of the potter's wheel is likely to have been a cumulative process developed over time in the city state workshops of the Near East. It appears from the available evidence, that the updraught kiln, potter's wheel, and workshop developed almost simultaneously across the Near East as suitable social and economic conditions were in place in order to foster its use. During practical experiments making 63 V-shaped bowls, Courty and Roux (1995, p. 750) noted that when rotating the wheel whilst adding the coils and finishing the rim, the act of producing pottery becomes more mechanised, and therefore speed of production increased. If the potter's wheel was invented in the Near East, as the evidence in Chapter 4 has suggested, how the technology of the potter's wheel was transferred to neighbouring Egypt needed next to be investigated.

The notion that craftspeople often do not have many technological choices and are limited by the resources and machinery available and by craft production traditions already available can only be supported until a change in technology can be viewed as beneficial for the society. This could be argued for the development of the potter's wheel, as it met the elite members of society's new requirements for their funerary and ritual pottery needs. This has disproved the most commonly held assumptions regarding the advent of potter's wheel, that it was created for standardisation and mass-production of vessels. This does not seem to be the case. It was initially created to furnish the elites with ritual and funerary vessels, elaborately manufactured as part of their increased luxury lifestyles. The use of potter's wheels was seemingly strongly controlled by elite temple personnel, who would also have guarded who would have had access to the vessels being produced by the potters. The Egyptians would have been able to easily adapt to the Near Eastern model. Egypt was unified under one leader, the Pharaoh, whose court would have controlled its craftspeoples' access to resources by attaching artisans' workshops to temples and estates. This would have meant that any new machinery introduced to Egypt would have needed the financial backing of an elite sponsor in order for it to be implemented.

Chapter 4 examined the reasons for inventing a technology and the significance of technological precursors for the potter's wheel in both Egypt and the Near East have been considered. Such evidence points to the premise that the Egyptians would have had the tools and the technology available to construct a set of potter's wheel bearings. They already had an extensive basalt vessel production programme in place, from which the wheel bearings were constructed. Heretofore, basalt was used as an elite funerary material in the site of Hierakonpolis and in the various Old Kingdom examples. The potter's wheel and the Twist Reverse Twist drills were the amongst the first ancient machines which used the hardest and most elite stones for new purposes in the manufacturing process rather than the end product. The Egyptians had the bureaucratic administrative means of control and redistribution of resources in order to initiate production of basalt wheel bearings should they wish to do so. By using basalt, a prestigious stone usually used for statuary to create the wheel bearings and to be used for industrial processes, highlights the ritual contexts and prestige for the elites who sponsored its use. The potter's wheel would therefore have been imbued with ritual prestige in its own right, and the greater skill required for learning to use it would perhaps have created a specialist potter class, albeit perhaps lower than other craft workers as suggested by textual evidence. The royal court had long-standing trade routes with the cities of the Levant and the Near East, perhaps even some colonies in the region of Canaan (Brandl, 1992; Faltings, 1998b) and so would have had access to the pottery produced on the potter's wheel if it were traded. Strong diplomatic relations with the rulers of the city-states would have instigated the sharing of ideas as well as commodities and craft workers to teach the use of the new technology.

During the transition from the Naqada I to II, (c.3550-3450 B.C.) as discussed in Chapter 5, handmade pottery begins to become more homogenised in form and style, evidenced in the appearance of Petrie's rough ware initially in Upper Egypt and later in Lower Egypt. This demonstrates the early beginnings of standardisation and the mass-production of pottery prior to the use of the potter's wheel. Until the Naqada II period, pottery was exclusively made of alluvial clays (which Egyptologists refer to as Nile Silt, see Appendix II). These clays are easy to work, shape and fire, and are the most ubiquitous clay in Egypt as they can be relatively easily collected from all along the alluvial plain of the Nile. The beginnings of the use of Marl clays, mostly only available from the Ballas and Qena regions of Egypt, and their generally higher firing temperature perhaps meant that the Egyptians had to be altogether more organised in their pottery production. Kilns and workshops with specialist potters were able to work this new clay and produce new and more varied vessels, with finer pastes of fabric e.g. wavy handled jars.

Conclusions

Chapter 5 further investigated the close early links between Egypt and the Near East, and considered the evidence for colonies of Canaanites at Buto and Ma'adi in Egypt. There was evidence that Canaanite potters were living and working in the Lower Egyptian sites and using their potter's wheels in Egypt to produce V-rimmed vessels using local Nile clays (Brandl, 1992, pp. 367-9). Pottery vessels from the Uruk region could also be found at Buto, such as holemouth jars, V-shaped bowls, and piecrust rims, and represented one third of the ceramic types at Buto, similar in nature the ceramic corpora of the Beersheva and Ma'adi regions (Faltings & Köhler, 1996, Abb 7.1; Köhler 1998, Tafel 74.1-2; Rizkana & Seeher, 1987, pg. 47). The potter's wheel it appears was first used in the Near East between 4000-4500 B.C. There is evidence for the potter's wheel in the Egyptian delta c.3500-3300 B.C. as Canaanite potters living in Egypt used a potter's wheel to thin and shape Canaanite style pottery but using local Egyptian Nile silt clays. As in the Near East, the instigation of the potter's wheel was through elite sponsorship, possibly through transference of potters between the royal courts, but more likely through colonisation of Canaan and Palestine; and through trade links between Egypt and these neighbours.

An analysis of where wheel thrown pottery occurs (whether that be in domestic, funerary or cultic contexts) and where pottery workshops have been located has aided recognition of the development of the kiln, the potter's wheel and the pottery workshop as a potentially elite-sponsored craft undertaken for a specific purpose other than the mass-production of domestic wares. The potter's wheel was used during Egypt's Middle Kingdom (c.2025-1700 B.C.) to manufacture mass-produced wheel made pottery. However, initially the potter's wheel was initially used to produce a select range of miniature and model vessels within particular context, of funerary and cultic offerings (Bárta, 1995, pp. 22-4). Several thousand of these miniature vessels have been found particularly in contexts such as the pyramid temples Sneferu at Meidum (Allen, 2006, pp. 19-21), Menkaure at Giza (Reisner, 1931, p. 228) and mastabas such as Ptahshepses at Abu Sir (Charvát, 1981, p. 148).

The techniques involved in using a potter's wheel are entirely different from those used in hand-throwing and may take a long time to learn as was addressed in Chapter 6. It is likely that the practical skills involved were passed down to the next generation, thus creating a specialised class similar in effect to the medieval guilds systems. It is arguably easier to learn to create a pot using coil, slab, pinching or paddle and anvil techniques, but it is something altogether different to use a potter's wheel. Firstly, the potter does not rely solely on their hands as the main shaping force, but instead utilises centrifugal force. Secondly, the potter is using a machine that probably had to be made by another craft worker i.e. the stonemason, with the knowledge of shaping and forming hard stones such as basalt, which is the stone which potter's wheels were commonly made from. In so doing, the stonemason was constructing the wheel to set specifications using their own tools and learned techniques.

Having identified suitable early wheel thrown pots from the Old Kingdom in the pottery collections of the Ashmolean Museum, Oxford Petrie Museum of Egyptian Archaeology, UCL and Egyptian Museum in Cairo, they were then examined for any manufacturing marks. Such marks, (e.g. the spirals, scrapes and striations) left by the ancient potters were compared to the marks created when using the replica Egyptian potter's wheel. These marks appeared to match with the pots dating to the 4th Dynasty that were being fashioned on the wheel, confirming Bárta's (1995) original thoughts that miniature vessels were wheel thrown, apart from the underside of the base, and are likely to be the first pots to be thrown on a potter's wheel, crucially, having centrifugal force being induced in order for throwing to be taking place. This has led to an increase in understanding and knowledge about the methods used to create such pots in ancient times. In later experiments, it was discovered that if the wheel was continuously rotated and a piece of string was applied to the base of the pot to cut it off from the hump, similar marks could be achieved on the base as noted on the archaeological examples. From these experiments, a list of manufacturing marks left on wheelthrown pottery has been created (see Chapter 6 and Tables 6.2 and 6.3) which will be useful for future identification of wheel- thrown pottery.

From the experiments undertaken in Chapter 6, the terminological problem of what constitutes a vessel thrown on a hand-spun potter's wheel and that which has been formed by coiling can now be resolved. Vessels which exhibit evidence of partial rotation on the rims, such as wavy handled jars or other storage jars, have not been thrown on a potter's wheel, but instead are likely to have been placed on an unmovable block or support such as a mat and then the pot rotated by the potter (as noted in the depiction of the pottery workshop tomb of Ty (Steindorff 1913, pp. 83-84; Épron and Daumas 1939, pp. 12, pl 71). Issues relating to the speed that a potter's wheel needs to achieve before it can be considered a "fast" versus a "slow" wheel can now be disputed. The author can find no such distinction, as the replica potter's wheel was successfully able to create thrown pottery at speeds lower than the suggested 50-150 r.p.m. (Rye 1981, p. 74), and even at the speed of 20 r.p.m., not considered by Jacobs and Edwards (1986, pp. 49-56) to be throwing. It is suggested that such terms as *fast* and *slow* wheel needs to be readdressed, if they should exist as a distinction at all.

As investigated in Chapter 7, the use of basalt to construct a wheel could indicate close involvement of the elites as it was difficult to procure and hew stone. Basalt was often sourced in hazardous desert conditions and required much organisation of personnel in order to procure it. Prior to its use for potter's wheels, basalt occurs only in highly prestigious contexts, usually for royal or elite funerary equipment in the form of vases, boundary or tomb marking *stelae* or statues. The use of basalt for creating basalt wheel bearings perhaps could signify wider changes within the fabric of Egyptian society. This book has tried to understand these changes by investigating who was determining the use of the potter's wheel in the first place and why it came to be invented at all. Technologies such as the potter's wheel and the twist reverse drill could be viewed as forms of control created by the newly established elite classes demonstrating their power and perhaps dominion over others. It signifies close technological and trade links to foreign societies such as those in Canaan, Palestine, and Mesopotamia. It seems from this book, that the potter's wheel, kiln, and workshop come together as an industrial package when the potter's wheel is adopted by the Egyptian court under Pharaoh Sneferu in the 4th dynasty. Sneferu possibly wanted to take advantage of the technology to manufacture vast quantities of miniature vessels to be used to serve his *ka* daily offerings of bread and beer as part of the *htp di nsw* formulae (Bárta, 1995, pp. 15-24).

CD7 and Meidum bowls possibly hold great significance for the second stage in the use of the potter's wheel, its spread to the general Egyptian populace. Meidum bowls are thought to have been utilised in the communal eating and serving of food, but also as an elite luxury tableware (Hendrickx, Op de Beeck, Raue, & Michiels, 2002, p. 277). They are regularly depicted in tomb and temple scenes associated with eating, food presentation, and feasting, and are often placed on pottery stands covered with a lid made of basketry (Wodzińska, 2006, p. 411). CD7 bowls by contrast, only occur in limited contexts, (1) The 4th dynasty Menkaure pyramid tomb builders' village at Giza (2) Old Kingdom Bakery area at Sheikh Said South (Vereecken, 2011, pp. 285, fig 9; Willems, *et al.*, 2009). CD7 bowls are unusual in that they are made of fine and medium fine Nile silt clay covered with a white wash rather than a red slip as the Meidum bowls have, and are scraped at the bases. Slipping vessels usually renders them impermeable to water, but washes are added after firing, so it is unlikely that the CD7 bowls were used for liquids. Many of these bowls apparently show clear signs of being rotated on the wheel as there are concentric striation marks on many of the rims and shoulders of the vessel, and the base is often irregular and trimmed (Wodzińska, 2006, pp. 405-429).

However, subsequent inspection of digital close-up photographs of the vessels by the author and Wodzińska suggest that this may not be the case. The CD7 vessels do not seem to exhibit the spiral striations, base cutting marks or the other characteristic marks associated with wheel throwing identified in Chapter 6. Further work must be undertaken on Meidum and CD7 bowls in order to ascertain when the pottery products of the potter's wheel became more widespread. Certainly, by the 6th dynasty many jars, bowls and plates of both open and closed forms are made on the potter's wheel. Exactly when this transistion occured, needs future investigation, but is at present, beyond the scope of this current work. Another question to be answered is how and when the kick wheel came to Egypt, and under what conditions.

Summary

Although the ultimate use of the potter's wheel's was for the mass production of pottery for varied and wide ranging functions, it seems that the initial use was much more specific. Its early use in Egypt was to create somewhat crudely made miniature vessels destined for quite lofty purposes, such as foundation deposits for pyramid mortuary temples, food offerings in mastabas and private offering chapels. These vessels are generally associated with offerings of the cult of the deceased person, often a king or member of the kings' court, and were produced in quite large numbers. Minature vessels often rather crudely fashioned out of Nile B2 silts but represented a completely new way of making and creating vessels by utilising centrifugal force. It may have been that the utilisation of this new technology of a potter's wheel using the prestigous basalt stone socket and pivots may have been more significant, as it allowed the elites to maintain their elite status.

Basalt had previously only been used to create eminent items such as sculpture, funerary and temple offering vessels and *stelae,* and these basalt items were usually only within the grasp of the elites. Initially, the potter's wheel may have been used to continue this trend and produce vessels for the elite workers and state officials who needed to have their funerary cults served with token food offerings using the eminent basalt wheel. The potter's wheel would have to have been constructed by a stonemason used to cutting and shaping hard stone such as basalt, no mean task for these workers when they only have copper chisels and stone drills and borers at their disposal. This perhaps indicates close links between the early craftsmen involved in the process, and a mutual understanding of what was required and the skills and techniques needed to produce it.

By utilising a new technology to create pots for funerary and cultic spheres on the highly prestigious basalt stone the elites were able to control what the specialist potters were creating and maintain a demand for similar items for the burgeoning administrative class that emerged during the Old Kingdom and who also required funerary items for their own tombs. The potter's wheel seems to mark fundamental changes within Egyptian society. Its use in Egypt indicates close contacts between Mesopotamia and the Levant, its importance as an early technological trade for use in cultic contexts, and the unusual step that Egyptians were often afraid to make i.e. changing a technology that already had been perfected. The Egyptians had already been creating remarkable pottery vessels, successfully hand-making, firing and usage long before the potter's wheel came to be invented. The use of the potter's wheel seems to be of much greater significant than simply a change in pottery styles and functions. It represents a different way of thinking for the Egyptians, that of moving beyond the domestic sphere and thinking on a broader, industrial workshop scale utilising a new technology in order to do so.

As well as being the first to utilised the potter's wheel for throwing, Egypt also contains the first depiction of a pottery kiln and a pottery workshop in a tomb, that of Ty at Saqqara (Épron & Daumas, 1939, p. 71; Steindorff, 1913, p. 83 and 84). The Egyptians went one technological step further than their Levantine neighbours, using the potter's wheel to produce thrown pottery created by inducing centrifugal force. The fact that such a large structure as a kiln was needed suggests that pottery production became a more industrialised process, with permanent workshops and specialised workers i.e. the potters were required to work all day every day solely to produce pots. There was clearly a demand beyond domestic household requirements that needed to be met. The use of the potter's wheel may have been fundamental in this process.

Through a thorough analysis of all available sources: manufacturing marks on pottery, provenanced potter's wheels, and depictions of potters in writings, in paintings and statues as illustrated in this book, can the origins of the potter's wheel begin to be understood. Through examining manufacturing marks on pottery and determining characteristics of wheel made marks by comparing them to experimental examples it is hoped a more complete view of when and in what manner the Egyptians were manufacturing their pottery vessels on the wheel has been gained. This book has sought to argue that the impact of the introduction of the potter's wheel to Egypt could not just have affected the Egyptian potters themselves through learning a new skill but also signalled the beginnings of a more complex and technologically advanced state.

References

ADAMS, B. (1974). *Ancient Hierakonpolis.* Warminster: Aris & Phillips Modern Egyptology series; v. 2.

ADAMS, B. (2000). *Excavations in the Locality 6 Cemetery at Hierakonpolis 1979-1985.* Oxford: British Archaeological Reports Int. Ser. 903.

ADAMS, B., & FRIEDMAN, R. F. (1992). Imports and Influences in the Predynastic and Protodynastic Settlement and Funerary Assemblages at Hierakonpolis. In E. C. van der Brink (Ed.), *The Nile Delta in Transition; 4th-3rd millennium B.C.* (pp. 317-388). Tell Aviv: Israel Exploration Society.

ADAMS, R. M., & NISSEN, H. J. (1972). *The Uruk Countryside. The Ideal Setting of Urban Societies.* Chicago: University of Chicago Press.

ADAMS, W. (1977). *Nubia, Corridor to Africa.* London: Allen Lane.

AKSAMIT, J. (1992). Petrie's Type D 46D and Remarks on the Production and Decoration of Predynastic Decorated Pottery. *Cahiers Ceramique d'Egyptienne*, 3, 17-21, pl 1-3.

ALDRED, C. (1980). *Egyptian Art in the Days of the Pharaohs 3100-320 B.C.* London: Thames and Hudson.

ALLEN, S. (2006). Miniature and model vessels in Ancient Egypt. In M. Barta (Ed.), *The Old Kingdom Art and Archaeology* (pp. 19-24). Prague: Proceedings of the conference held in Prague, May 31-4 June 2004. Acedemy of Sciences for the Czech Rep.

AMER, M. (1933). Annual Report of the Maadi excavations 1930-32. *Bulletin of the Faculty of Arts*, I.

AMER, M. (1936). Annual Report of the Maadi excavations 1935. *Chronique d'Égypte, XI*.

AMIRAN, R. (1963). *Ancient Pottery of the Holy Land.* Jerusalem: Ramat Gan (in Hebrew).

AMIRAN, R. (1978). *Early Arad. The Chalcolithic Settlement and Early Bronze Age City. Vol I. The First-Fifth Seasons of Excavation 1962-1966.* Jerusalem: The Israel Exploration Society.

AMIRAN, R., BEIT-AUCH, Y., & GLASS, J. (1973). Interrelationships between Arad and sites in Southern Sinai in the early Bronze Age. *Israel Exploration Journal*, 23 (4), 193-197.

AMIRAN, R., & SHENDOV, D. (1966). Experiments in the Working of Ancient Potter's wheels and Producing Pottery with them. *Bulletin JPES*, 30, 85-87.

AMIRAN, R., & SHENDOV, D. (1984). Experiments with an Ancient Potter's Wheel. In P. Rice (Ed.), *Pots and Potters. Current approaches in Ceramic Archaeology* (pp. 107-112). IoA Monographs, University of California.

ANTHES, R. (1965). *Mit Raahineh, 1956.* Philadelphia.

ARNOLD, D. E. (1985). *Ceramic theory and cultural process.* Cambridge: Cambridge University Press.

ARNOLD, D. E. (2000). Does the Standardization of Ceramic Paste Really Mean Specialization? *Journal of Archaeological Method and Theory*, 7, 333-375.

ARNOLD, Di. (1981). *Der Tempel des Königs Montuhotep von Deir el Bahri. Band III: Die königlichen Beigaben.* Mainz.

ARNOLD, Do. (1976). Wandbild und Scherbennefund. Zur Töpfertechnik der alter Ägypter von Beginn der pharaonischen Zeit bis zu den Hyksos. *Mitteilungen des Deutchen Institutes Abteilung Kairo*, 32, 1-34.

ARNOLD, Do. (1993). Fascicle 1:Techniques and Traditions of Manufacture in the Pottery of Ancient Egypt. In Do. Arnold, & J. Bourriau (Eds.), *An Introduction to Ancient Egyptian Pottery* (pp. 9-141). Mainz am Rhein: Verlag Philipp von Zabern.

ARNOLD, Do; BOURRIAU, J. (Eds.), *An Introduction to Ancient Egyptian Pottery* Mainz am Rhein: Verlag Philipp von Zabern.

ASSMAN, J. (1996). *The Mind of Egypt. History and Meaning in the time of the Pharaohs.* New York: Metropolitan Books, Henry Holt and Co.

ASTON, B. G. (1994). *Ancient Egyptian Stone Vessels. Materials and Forms.* Cairo, Heildelberg: Universität Heidelberg, Heidelberger Orientverlag, Studien zur Archäologie und Geschichte Altägyptens 5.

BABA, M. (2005). Understanding the HK Potters: Experimental Firings. *Nekhen News*, 17, 20-1.

BABA, M. (2006). The pottery Kilns at HK11C Revisited. *Nekhen News*, 18, 19.

BAGH, T. (2011). *Finds from W. M. F. Petrie's Excavations in Egypt in the Ny Carlsberg Glyptotek.* Copenhagen: Ny Carlsberg Glyptotek.

BAINES, J. (1983). Literacy and Ancient Egyptian Society. *Man*, 18 (3), 572-599.

BAINES, J. (1994). On the Status and Purpose of Ancient Egyptian Art. *Cambridge Archaeological Journal*, 4 (1), 67-94.

BAINES, J. (2007). *Visual and Written Culture in Ancient Egypt.* Oxford: Oxford University Press.

BAINES, J., & LACOVARA, P. (2002). Burial and the dead in ancient Egyptian society. *Journal of Social Archaeology* (1), 5-36.

BAINES, J., & YOFFEE, N. (1998). Order, Legitimacy and Wealth in Ancient Egypt and Mesopotamia. In G. M. Feinman, & J. Marcus (Eds.), *Archaic States: A*

Comparative Perspective (pp. 199-260). Santa Fe: School of American Research Press.

BALLET, P. (1987). Essai de Classification des Coupes Type Maidum-Bowl du Sondage de Ayn-Asil (Oasis de Dakhla) Tpologie et Evolution. *Cahiers de la Ceramique Égyptienne, 1*, 1-15.

BARD, K. A. (1988). A quantitative analysis of the Predynastic Burials in Armant Cemetery 1400-1500. *Journal of Egyptian Archaeology*, 39-55.

BARD, K. A. (1994). *From Farmers to Pharaohs: Mortuary Evidence for the Rise of Complex Society in Egypt.* Sheffield.

BARD, K. A. (2000). The Emergence of the Egyptian State. In I. Shaw (Ed.), *The Oxford History of Ancient Egypt* (pp. 61-88). Oxford: Oxford University Press.

BARLEY, N. (1994). *Smashing Pots: Feats of Clay in Africa.* London: British Museum Press.

BÁRTA, M. (1995). Pottery Inventory and the Beginning of the IVth Dynasty. *Gottinger Mitteilungen, 149*, 15-24.

BÁRTA, M. (2001). *Abusir V: The Cemeteries of Abusir South I.* Prague: Czech Institute of Egyptology.

BÁRTA, M. (2005). Location of Old Kingdom Pyramids in Egypt. *Cambridge Archaeological Journal, 15*, 177-191.

BASMACHI, F. (1947). The Votive Vase from Warka. *Sumer, 3*, 118-27.

BEALE, T. W. (1978). Bevelled Rim Bowls and their Implications for the Change and Economic Organization in the Later Fourth Millennium B.C. *Journal of Near Eastern Studies, 37* (4), 289-313.

BENDERITTER, T, HIRST, J. J (2012) http://www.osirisnet.net/e_centra.htm [accessed 24th August 2012]

BERG, I. (2007). Meaning in the making: The potter's wheel at Phylakopi, Melos (Greece). *Journal of Anthropological Archaeology, 26*, 234-252.

BINFORD, L. (1965). Archaeological Systematics and the Study of Culture Process. *American Antiquity, 31* (2), 203-10.

BIRKS, T. (1979). *Pottery. A Complete Guide to Pottery Making Techniques.* London: Alphabooks, A & C Black.

BLACKMAN, M. J., STEIN, G. L., & VANDIVER, P. B. (1993). The Standardisation Hypothesis and Ceramic Mass Production: Technological, Compositional, and Metric Indexes of Craft Specialization at Tell Leilan, Syria. *American Archaeology, 58* (1), 60-80.

BLACKMAN, W. (1927). *The Fellahin of Upper Egypt.* London: Harrap Press; AUC Press.

BOATRIGHT, D., & HODGKINSON, A. (2010). The Kiln Excavation. In *Report to the SCA on archaeological survey and excavation undertaken at Medinet el Gurob 4-15 April 2010* (pp. 13-16). unpublished report.

BORCHARDT, L. (1932). Ein Brot. *Zeitschrift fur Agyptische Sprache und Altertumskunde, 68*, 73-79.

BOURRIAU, J. (1981). *Umm el Ga'ab. Pottery from the Nile Valley Before the Arab Conquest.* Cambridge: Fitzwilliam Museum, Cambridge University Press.

BOURRIAU, J. (2002). The Beginnings of Amphorae Production in Egypt. In J. Bourriau, & J. Phillips (Eds.), *Invention and Innovation. The Social Context of Technological Change 2. Egypt, the Aegean, and the Near East* (pp. 78-95). London: Oxbow Books.

BOURRIAU, J., NICHOLSON, P., & ROSE, P. (2000). Pottery. In P. S. Nicholson (Ed.), *Ancient Egyptian Materials and Technology* (pp. 121-47). Cambridge: Cambridge University Press.

BRAIDWOOD, R. J., & HOWE, B. (1960). *Prehistoric Investigations in Iraqi Kurdistan.* Chicago: University of Chicago Press.

BRANDL, B. (1989). Observations on the Early Bronze Age Strata of Tell Erani. In P. de Miroschedjii (Ed.), *L'urbanisation de la Palestine à l'Age du Bronze Ancien. Bilan et perspectives des reserches actuelles* (pp. 357-388). Actes du Colloque d'Emmais (20-24 octobre 1986) BAR International Series 527 (ii).

BRANDL, B. (1992). Evidence for Egyptian Colonisation in the Southern Coastal Plain and Lowlands of Canaan during the EB I period. In E. C. van der Brink (Ed.), *The Nile Delta in Transition 4th-3rd millennium B.C.* (pp. 441-448). Tel Aviv: Israel Exploration Society.

BRAUN, E. (2003). South Levantine Encounters with Ancient Egypt at the beginning of the Third Millennium. In R. Matthews, & C. Roemer (Eds.), *Ancient Perspectives on Ancient Egypt* (pp. 21-37). London: UCL Press.

BREASTED, J. H. (1948). *Servant Statues.* Washington: Pantheon Books. The Bollingden Series.

BREWER, D. J., REDFORD, D. B., & REDFORD, S. (1994). *Domestic plants and animals: the Egyptian origins.* Warminster: Aris & Phillips.

BRIL, B., LEDEBT, A., DIETRICH, G., & ROBY-BRAMI, A. (1998). *Perception-Action Coupling and the Planning of Action.* Paris: EDK.

BROUWER, J. (1987). An Exploration of the Traditional Division of Labour Between the Sexes in South Indian Craft. In A. Menefee Singh, & A. Kelles-Viitanen (Eds.), *Invisible Hands: Women in Home-based Production* (pp. 145-163). London: Sage.

BRUMFIEL, E. M., & EARLE, T. K. (1987). *Specialization, Exchange, and Complex Societies.* Cambridge: Cambridge University Press.

BRUNNER, H. (1964). Gebert des Gotteskönigs. *Ägyptologische Abhandlungen.*

CARDEW, M. (2002). *Pioneer Pottery.* London, Ohio: A & C Black, American Ceramic Society.

CHARVÁT, P. (1981). *The Mastaba of Ptahshepses. The Pottery.* Praha: Univerzita Karlova.

CHAZAN, M., & LEHNER, M. (1990). An ancient analogy: Pot baked bread in ancient Egypt and Mesopotamia. *Paléorient, 16* (2), 21-35.

CHERPION, N. (1989). *Mastabas et hypogeés de l'Ancien Empire. Le problème de la datation.* Bruxelles: Connaissance de l'Égypte anciennes de l'Ancien Empire.

CHILDE, V. G. (1954). Rotary Motion. In C. Singer, E. J. Halmyard, & A. R. Hall (Eds.), *A History of Technology from early times to fall of Ancient Empires* (pp. 187-215). Oxford: Clarendon Press.

CHILDE, V. G. (1957). The Bronze Age. *Past and Present, 12* (1), 2-15.

CLAYTON, P. (1994). *Chronicle of the Pharaohs.* New York: Thames and Hudson.

COLBECK, J. (1982). *Pottery: The Technique of Throwing.* London & Sydney: B. T. Batsford Limited.

COMMENGE-PELLERIN, C. (1987). *La Poterie d'Abou Matar et de l'Ouadi Zoumeili (Beersheva) au IVe millenaire avant l'Ere Chretienne.* Paris: Association Paleorient (Les Cahiers du Centre de Researche Francais de Jerusalem no. 3).

COMMENGE-PELLERIN, C. (1990). *La Poterie de Safadi (Beersheva) au IVe millenaire avant l'Ere Chretienne.* Paris: Association Paleorient (Les Cahiers du Centre de researche Francais de Jerusalem, no. 5).

COSTIN, C. L. (1991). Craft Specialisation: Issues in Defining, Documenting and Explaining the Organizations of Production. In M. B. Schiffer (Ed.), *Archaeological Method and Theory* (Vol. 3, pp. 1-56). Tuscon: University of Arizona Press.

COSTIN, C. L. (1996). Exploring the Relationship Between Gender and Craft in Complex Societies: Methodological and Theoretical Issues of Gender Attribution. In R. P. Wright (Ed.), *Gender and Archaeology* (pp. 111-140). Philadelphia: University of Pennslyvania Press.

COULSON, W. D., & LEONARD, A. J. (1983). The Naukratis Project. *Muse, 17,* 64-71.

COURTY, M. A., & ROUX, V. (1995). Identification of Wheel Throwing on the Basis of Ceramic Surface Features and Microfabrics. *Journal of Archaeological Science, 22,* 17-50.

CREWE, L. (2007). Sophistication in Simplicity: The First Production of Wheelmade Pottery on Late Bronze Age Cyprus. *Journal of Mediterranean Archaeology, 20* (2), 209-238.

CURTIS, F. (1962). The Utility Pottery Industry of Bailen, Southern Spain. *American Anthropology, 64,* 486-503.

D'AURIA, S., LACOVARA, P., & ROEHRIG, C. H. (1988). *Mummies and Magic. The Funerary Arts of Ancient Egypt.* Museum of Fine Arts, Boston, Dallas Museum of Art.

DAVEY, C. (1979). Some ancient Near East pot bellows. *Levant, 11,* 10-11.

DAVID, R. (2003). *Handbook to Life in Ancient Egypt* (Revised ed.). New York: Facts on File Inc.

DAVIES, N. DE. (1918). *The Egyptian Expedition 1916-1917.* New York: Supplement to the Bulletin of the Metropolitan Museum of Art.

DAVIES, N. DE.G. (1930). *The Tomb of Ken-Amun at Thebes Vols I-II.* New York.

DAVIES, N. DE G. (1953). *The Temple of Hibis in Khargeh Oasis. Part III* (Vol. XVII). New York: Publications of the Metropolitan Museum of Art Egyptian Expedition.

DAVIES, V. (2004). Hatshepsut's Use of Tuthmosis III in Her Program of Legitimation. *Journal of the American Research Center in Egypt, 41,* 55-66.

DEBONO, F., & MORTENSEN, B. (1990). *El Omari: a Neolithic settlement and other sites in the vicinity of Wadi Hof, Helwan with appendixes on geology.* Mainz.

DESSEL, J. P. (2009). *Lahav I. Pottery and Politics: The Halif terrace site 101 and Egypt in the 4th millennium B.C.E.* Winona Lake, Indiana: Eisenbrauns.

DESSEL, J. P., & JOFFE, A. H. (2000). Alternative Approaches to Early Bronze Age Pottery. In G. Philip, & D. Baird (Eds.), *Ceramics and Change in the Early Bronze Age of the Southern Levant* (pp. 31-48). Sheffield: Sheffield Academic Press.

DOBRES, M.-A. (1999). Technology's Links and Chaines: The Processual Unfolding of Technique and Technician. In M. A. Dobres, & C. R. Hoffman (Eds.), *The Social Dynamics of Technology* (pp. 124-161). Washington & London: Smithsonian Institute Press.

DOBRES, M.-A. (2000). *Technology and Social Agency.* Oxford: Blackwell Publishers.

DODSON, A. (1995). *Monarchs of the Nile.* Cairo: American University in Cairo Press.

DOHERTY, S. K. (2014). *Potter's Wheel Videos.* Available: https://www.youtube.com/user/Ramessesmissy/feed.

DOHERTY, S. K. (in press). Scrapes, Strings and Striations: Replicating Ancient Egyptian Wheelthrown Pottery. In T. Rzeuska, & A. Wodzińska (Eds.), *Old Kingdom Pottery Workshop: Chapter 2.* Warsaw.

DORMAN, P. F. (2002). *Faces in Clay. Techniques, Imagery and Allusion in a corpus of ceramic sculpture from Ancient Egypt.* Mainz, Rhein: Verlag Philipp von Zabern.

DOTHAN, M. (1959). Excavations at Meser. Preliminary Report on the second season. *Israel Exploration Journal, 9,* 13-29.

DRENKHAHN, R. (1976). *Die Handwerker und ihre Tätigkeiten im alten Ägypten (Ägyptologische Abhandlungen).* Berlin: Harrassowitz.

DUISTERMAAT, K. (2008). *The Pots and Potters of Assyria. Technology and organisation of production, ceramic sequence and vessel function at Late Bronze Age Tell Sabi Abyad, Syria.* Turnhout, Belgium: Brepols Publishers, n.v.

EARLE, T. K. (1987). Specialization and the production of wealth: Hawaiian chiefdoms and the Inka Empire. In E. M. Brumfield, & T. K. Earle (Eds.), *Specialization, Exchange and Complex Societies* (pp. 64-75). Cambridge: Cambridge University Press.

EKHOLM, K., & FRIEDMAN, J. (1982). "Capital" Imperialism and Exploitation in Ancient World-Systems. *Review*, 6 (1), 87-109.

EDWARDS, I. E. (1986). *The Pyramids of Egypt* (1947 reprint ed.). New York and London: Viking Penguin Inc.

EDWARDS, I., & JACOBS, L. (1986). Experiments with Stone "Pottery wheel" bearings- notes on the use rotation in the production of ancient pottery. *Newsletter, Department of Pottery Technology*, 4, 49-56.

EDWARDS, I., & JACOBS, L. (1987). Experiments with Stone "Pottery wheel" bearings- notes on the use rotation in the production of ancient pottery. *Newsletter, Department of Pottery Technology*, 4, 49-55.

EL-KHOULI, A. (1978). *Egyptian Stone Vessels, Predynastic Period to Dynasty III* (Vol. II). Mainz: Philipp von Zabern.

EL-KHOULI, A. (1991). *Meidum*. (G. T. Martin, Ed.) The Australian Centre for Egyptology: Reports 3.

EMERY, W. B. (1962). *A Funerary Repast in an Egyptian Tomb of the Archaic Period*. Leiden: Nederlands Instituut voor het Nabije Oosten.

EMERY, W. B. (1963). Preliminary Report on the Excavations at Buhen. *Kush*, 2, 116-120.

EMPEROUR, J.-Y., & PICON, M. (1992). La Reconnaissance des Productions des Ateliers Céramiques: L'Exemple de La Maréotide. In P. Ballet (Ed.), *Cahiers de la Céramique Égyptienne 3. Ateliers de Potiers et Productions Céramiques en Égypte* (pp. 145-152). Le Caire: Insituts Français D'Archéologie Orientale du Caire.

ENGBERG, R. M., & SHIPTON, G. M. (1934). *Notes on the Chalcolithic and Early Bronze Age Pottery of Megiddo*. Chicago.

ÉPRON, L., & DAUMAS, F. (1939). *Le Tombeau de Ti Fasc.1, Les approches de la chapelle*. Cairo: Le Caire: Impr. de l'Institut français d'archéologie orientale.

ERICSSON, K. A., & LEHMANN, A. C. (1996). Expert and exceptional performance: Evidence from maximal adaptation to task restraints. *Annual Review of Psychology*, 47, 273-305.

EVELY, D. (1988). The Potter's Wheel in Minoan Crete. *Annual Bulletin School of Athens*, 83, 83-126.

EVELY, D. (2000). *Minoan Crafts: Tools and Techniques, vol 2*. Jonsered: Paul Åström Förlag.

FAKHRY, A. (1961). *The Monuments of Sneferu at Dashur. Vol. II: The Valley Temple. Part II: The Finds*. Cairo: General Organisation for Government Printing Offices.

FALTINGS, D. (1989). Die Keramik aus den Grabungen an der nordlichen Pyramide des Snofru in Dahschur, Arbeitsbericht uber die Kampagnen 1983-1986. *Mitteilungen des Deutschen Archäologischen Instituts Abteilung Kairo (MDAIK)*, 45, 133-154.

FALTINGS, D. (1998a). Canaanites at Buto in the early fourth millennium B.C. *Egyptian Archaeology*, 29-32.

FALTINGS, D. (1998b). Recent Excavations in Tell el-Fara'in/ Buto: New Finds and Their Chronological Implications. In C. J. Eyre (Ed.), *Proceedings of the 7th International Congress of Egyptologists. Cambridge 3-9 September 1995* (pp. 365-375). Leuven: Peeters. Orientalia Lovaniensia Analecta.

FALTINGS, D., & KÖHLER, E. C. (1996). Vorbericht über die Ausgrabungen des DAI in Tell el Fara'in Buto. *Mitteilungen des Deutschen Archäologischen Instituts, Abteilung Kairo (MDAIK)*, 52, 87-114.

FAULKNER, R. O. (1952). The Stela of the Master-Sculptor Shen. *The Journal of Egyptian Archaeology*, 38, 3-5.

FAULKNER, R. O. (1969). *The Ancient Egyptian Pyramid Texts*. Oxford: Clarendon Press.

FILIPOVIC, M. S. (1951). Primitive Ceramics Made by Women among the Balkan Peoples. *Siberian Academy of Sciences Monograph Ethnographic Institute*, 171 (2), 157-70.

FINNESTAD, R. B. (1985). *Image of the World and Symbol of the Creator*. Wiesbasden: O. Harrassowitz. Studies in Oriental Religions.

FIRTH, C. M., & GUNN, B. (1926). *Teti Pyramid Cemeteries* (Vols. I-II). Cairo.

FISCHER, H. G. (1989). Women in the Old Kingdom and Heracleopolitan Period. In B. S. Lesko (Ed.), *Women's Earliest Records, From Ancient Egypt and Western Asia* (pp. 5-43). Atlanta, Georgia: Brown Judaic Studies 166, Scholars Press.

FLENTYE, L. (2007). The Mastabas of Ankh-haf (G7510) and Akhethetep and Meretites (G7650) in the Eastern Cemetery at Giza: A Reassessment. In Z. A. Hawass, & J. Richards (Eds.), *The Archaeology and Art of Ancient Egypt. Essays in Honor of David B. O'Connor Vol I* (pp. 291-308). Annales du Services des Antiquities de l'Egypte Cahier no. 36.

FORBES, R. J. (1958). *Man the Maker. A History of Technology and Engineering*. London: Constable & Co Ltd.

FOSTER, G. M. (1959a). The Coyotepec molde and some associated problems of the potter's wheel. *Southwestern Journal of Anthropology*, 15, 53-63.

FOSTER, G. M. (1959b). The potter's wheel an analysis of an idea and artifact in invention. *Southwestern Journal of Anthropology*, 15 (2), 99-119.

FOSTER, G. M. (1967). *Tzintzuntzan: Mexican Peasants in a Changing World*. New York: Little Brown and Co.

FRANCHET, L. (1911). Instructions destinées aux Archéologues et Ethnographes dans le but de recueillir des renseignements relatifs à la technique céramique, verrière et métallurgique chez les peuples primitifs. *Extrait de L'Homme préhistorique*, 1.

FRANKEN, H. J. (1971). Analysis of Methods of Potmaking in Archaeology. *Harvard Theological Review*, 64, 227-255.

FRANKEN, H. J. (2005). *A History of Potters and Pottery in Ancient Jerusalem*. London: Equinox.

FRANKEN, H. J., & KALSBEEK, J. (1975). *The Potters of a Medieval Village in the Jordan Valley*. Amsterdam, North Holland.

FRANKFORT, H. (1924). *Studies in Early Pottery of the Near East* (Vol. 2). London: Royal Anthropological Institute of Great Britain and Ireland 1924-1927.

FRANKFORT, H. (1941). The Origins of Monumental Architecture in Egypt. *American Journal of Semitic Languages and Literatures, 58*, 329-358.

FRANKFORT, H., & PENDLEBURY, J. D. (1933). *City of Akhenaten, Part II: The North Suburb and desert altars. Excavations at Tell el Amarna during seasons 1926-1932*. London: The Egypt Exploration Society, 40th Memoir.

FREDRICKSON, V.-A., & ELSASSER, A. B. (1972). *The Wepemnofret stela*. Berkeley: Robert H. Lowie Museum of Anthropology and University Art Museum, University of California.

FREESTONE, I., & GAIMSTER, D. (1997). *Pottery in the Making: World Ceramic Traditions*. London: BM Press.

FRIEDMAN, R. (1994). Aspects of Domestic Life and Religion. In L. Lesko, *Pharaohs Workers* (pp. 95-117). Ithaca.

FRIEDMAN, R. (1996). The Ceremonial Centre at Hierakonpolis Locality HK29A. In Spencer (Ed.), *Aspects of Early Egypt* (pp. 16-35). London: British Museum Press.

FRIEDMAN, R. (2004). Predynastic Kilns at HK11C: One side of the story. *Nekhen News, 16*, 18-19.

FRIEDMAN, R. (2006). The forgotten Potter of Horemkhawef. *Nekhen News, 18*, 20.

GABLEHOUSE, C. (1969). *Helicopters and Autogiros; A History of Rotating-wing and V/STOL Aviation*. Philadelphia: J.B. Lippincott Company.

GAHLIN, L. (2001). *Egypt. Gods, Myths and Religion*. London: Anness Publishing Ltd.

GALE, R., GASSON, P., HEPPER, N., & KILLEN, G. (2000). Wood. In P. T. Nicholson, & I. Shaw (Eds.), *Ancient Egyptian Materials and Industries* (pp. 334-371). Cambridge: Cambridge University Press.

GARDINER, A. H. (1909). *The Admonistions of an Egyptian Sage*. Leipzig: Hildesheim 1969 reprint.

GARDINER, A. H. (1937). Late Egyptian Miscellanies. *Bibliotheca Aegyptiaca VII*.

GARSTANG, J. (1902). A Pre-dynastic Pot Kiln recently discovered at Mahasna, Egypt. *Man, 2*, 38-40.

GATES, C. (2003). *Ancient Cities. The Archaeology of Urban Life in the Ancient Near East, Egypt, Greece and Rome*. London and New York: Routledge Press.

GAUTHIER, H. (1912). *Temple de Ouadi es-Sebua. Les Temples Immerges de la Nubie* (Vol. 5). Le Caire.

GELBERT, A. (1997). De l'Élaboration au Tour au Tournage sur Motte: Difficultés Motrices et Conceptuelles. *Techniques et Cultures, 30*, 1-23.

GELLER, J. R. (1984). *The Predynastic Ceramics Industry at Hierakonpolis, Egypt*. St Louis.

GERISCH, R. (2007). Appendix 3: Charcoal Remains. In P. T. Nicholson (Ed) *Brilliant Things for Akenaten. The Production of Glass, Vitreous Materials and Pottery at Amarna Site O45.1* (pp. 169-176). London: EES Excavation Memoir 80.

GHALY, H. (1992). Pottery Workshops of Saint Jeremia (Saqqara). In P. Ballet (Ed.), *Cahiers de la Céramique Égyptienne 3. Ateliers de Potiers et Productions Céramiques en Égypte* (pp. 161-171). Le Caire: Insituts Français D'Archéologie Orientale du Caire.

GILLINGS, R. J. (1982). *Mathematics in the time of the pharaohs*. New York, Dover, London: Constable.

GOEDICKE, H. (1979). Cult-Temple and "State" During the Old Kingdom in Egypt. In E. Lipiński (Ed.), *State and Temple Economy in the Ancient Near East* (Vol. I, pp. 113-132). Leuven: Departement Oriëntalistiek.

GOODY, J. (1971). *Technology, Tradition and the State in Africa*. London, Ibadan, Accra: Oxford University Press for the International African Institute.

GOPHNA, R. (1996). *Excavations at Tel Dalit*. Tel Aviv: Ramot Publications.

GOPHNA, R., & VAN DEN BRINK, E. C. (2002). Core-Periphery Interation between the Pristine Egyptian Naqada IIIb State, Late Early Bronze Age I Canaan, and Terminal A-Group Lower Nubia: More Data. In E. C. van den Brink, & T. E. Levy (Eds.), *Egypt and the Levant: interrelations from the 4th through the early 3rd millennium B.C.E* (pp. 280-281). London: Leicester University Press.

GOREN, Y. (1995). Shrines and Ceramics in Chalcolithic Israel: The View through the Petrographic Microscope. *Archaeometry, 37*, 287-305.

GOSSELAIN, O. P. (1999). In Pots We Trust. The Processing of Clay and Symbols in Sub-Saharan Africa. *Journal of Material Culture, 4* (2), 205-230.

GOSSELAIN, O. P. (1994). Skimming Through Potters' Agendas: An Ethnoarchaeological Study of Clay selection strategies in Cameroon. In *Society, Culture and Technology in Africa* (pp. 99-107). Philadelphia: MASCA Research Papers in Science and Archaeology, supplement to vol 11. University of Pennsylvania Museum of Archaeology and Anthropology.

GOULDER, J. (2010). Administrators' bread: an experiment-based re-assessment of the functional and cultural role of the Uruk bevel-rim bowl. *Antiquity, 84*, 351-362.

GRACE, R. (1989). *Interpreting the Function of Stone Tools: The quanitifcation and computerisation of microwear analysis*. B.A.R. International Series 474.

HANNIG, R. (2003). *Ägyptishches Wörterbuch I*. Mainz: Verlag Philipp von Zabern.

HARELL, J. A., & BROWN, T. M. (1995). An Old Kingdom Basalt Quarry at Widan el-Faras and the quarry road to lake Moeris. *Journal of the American Research Center in Egypt* (32), 71-91.

HARLAN, J. F. (1982). Excavations at locality IIC. In M. A. Hoffman, *The Predynastic of Hierakonpolis-An Interim Report* (pp. 14-25). Cairo and Macomb, Ill.

HARPUR, Y. (1987). *Decoration In Egyptian Tombs*. London and Leiden: Taylor & Francis.

HARPUR, Y. (2001). *The Tombs of Nefermaat and Rahotep at Maidum*. Oxford.

HARRISON, H. W. (1928). *Pots and Pans*. New York.

HASSAN, F. A. (1985). Radiocarbon chronology of Neolithic and Predynastic sites in Upper Egypt and the Delta. *African Archaeological Review*, 3 (1), 95-115.

HASSAN, S. (1948). *Excavations at Giza. 1934-1935* (Vol. VI Part II). Cairo: Services des Antiquities de l'Egypte.

HAYES, W. C. (1953). *The Scepter of Egypt* (Vol. I). New York: Metropolitan Museum of Art.

HELMS, M. (1993). *Craft and Kingly Ideal*. Austin: University of Texas Press.

HENDRICKX, S. (1996). The Relative Chronology of the Naqada Culture. In J. Spencer (Ed.), *Aspects of Early Egypt* (pp. 36- 69). London: British Museum Press.

HENDRICKX, S., FRIEDMAN, R., & LOYENS, F. (2000). Experimental Archaeology Concerning Black-topped Pottery from Ancient Egypt and the Sudan. *Cahiers Ceramique Egyptienne 6*, *3*, 171-188.

HENDRICKX, S., OP DE BEECK, L., RAUE, C., & MICHIELS, D. (2002). Milk, Beer and Bread Technology during the Early Dynastic Period. *Mitteilungen des Deutschen Archäologischen Instituts, Abteilung Kairo (MDAIK)*, *58*, 277-304.

HERBERT, E. (1993). *Iron, Gender and Power, Rituals of Transformation in African Societies*. Bloomington and Indianapolis: Indiana University Press.

HILL, J. N. (1977). *Explanation of Prehistoric Change*. Albuquerque: University of New Mexico Press (School of American Research Advanced Seminars).

HODGES, H. W. (1971). *Technology in the Ancient World*. New York.

HODGKINSON, A. (2012). The Excavation of "the Industrial Area": IA1. In I. Shaw (Ed.), *Report to the SCA on the archaeological survey and excavation undertaken at Medinet el-Gurob, 27 March-17th April 2012* (pp. 11-14). unpublished report.

HOFFMAN, M. A. (1982). *The predynastic of Hierakonpolis: an interim report*. Giza and Western Illinois, Cairo University Herbarium and the Dept of Sociology and Anthropology, Western Illinois University: Egyptian Studies Association, no. 1.

HOFFMEIR, J. (1993). The use of basalt floors of Old Kingdom pyramid temples. *The Journal for the American Research Center in Egypt*, 117-123.

HÖLSCHER, U. (1939). *The Temples of the Eighteenth Dynasty. The Excavations of Medinet Habu* (Vol. XLI). Chicago: Chicago Oriental Institute Publications.

HOLTHOER, R. (1977). *New Kingdom Pharaonic Sites: The Pottery*. London.

HOPE, C. A. (1979). Dakhleh Oasis Project- report on the study of the pottery and kilns. *Journal of the Society for the Study of Egyptian Antiquities*, 9 (4), 187-201.

HOPE, C. A.(1981). Two Ancient Egyptian Potter's Wheels. *Journal of the Society for the Study of Egyptian Antiquities*, *11*, 127-33.

HOPE, C. A. (1982). Concerning Egyptian Potter's wheels. *Journal of Egyptian Antiquities*, *12*, 13-14.

HOPE, C. A. (1987a). Experiments in the manufacture of Ancient Egyptian pottery. In W. I. Edwards, C. A. Hope, & E. R. Segnit (Eds.), *Ceramics from the Dakhleh Oasis* (pp. 103-105). Melbourne: Visctoria College Archaeological Research Unit occasional paper no. 1.

HOPE, C. A. (1987b). *Egyptian Pottery*. Aylesbury, Bucks: Shire Egyptology.

HOPE, C. A. (1995). Pottery Kilns from the Oasis at el-Dakla. In D. Arnold, & J. Bourriau, *An Introduction to Ancient Egyptian Pottery* (pp. 121-127). Mainz am Rhein: verlag Philip von Zabern.

HUGHES, M. J. (Ed.). (1981). *Scientific Studies in Ancient Ceramics*. London: British Museum Press.

JACOBS, P. F., & BOROWSKI, O. (1993). Notes and News Site 101: Chalcolithic Period. *Israel Exploration Journal*, *43*, 69-70.

JACQUET, J. (1965). The Architect's Report. In R. Anthes (Ed.), *Mit Rahineh* (pp. 45-59). Philadelphia: The University Museum, University of Pennsylvania.

JAMES, T. G. (1953). *The Mastaba of Khentika called Ikhekhi, Memoir of Archaeological Survey of Egypt, Memoir 13th*. London.

JANSSEN, J. (1975). Prolegomena to the Study of Egypt's Economic History during the New Kingdom. *Studien zur Altägyptischen Kultur*, *3*, 127-185.

JEFFREYS, D. G. (1998). The topography of Heliopolis and Memphis: some cognitive aspects. In H. Guksch, & D. Polz (Eds.), *Stationen: Beitrage zur Kulturgeschichte Agyptens* (pp. 63-71). Mainz: Rainer Stadelmann Gewimdet.

JEFFREYS, D. G. (2010). Regionality, Cultural and Cultic Landscapes. In W. Wendrich (Ed.), *Egyptian Archaeology* (pp. 102-118). Chichester: Wiley-Blackwell.

JOHNSTON, G. (1987). Nine thousand years of social change in western Iran. In F. Hole (Ed.), *The Archaeology of Western Iran*. Washington D. C.: Smithsonian Inst Press.

JOHNSTON, R. H. (1977). The Development of the Potter's Wheel: An Analytical and Synthesizing Study. In H. Lechtman, & R. Merrill (Eds.), *Material Culture*.

Styles, Organization and Dynamics of Technology (pp. 169-210). St Paul, New York, Boston, Los Angeles and San Francisco: Proceedings of the American Ethnological Society. West Publishing Company.

JUNKER, H. (1912). *Bericht über die Grabungen der Kiaserlichen Akademie der Wissenschaften in Wien auf dem Friedhof in Turah.* Wien, Friedhof.

JUNKER, H. (1929). *Gîza: Bericht über die von der Akademie der Wissenschaften in Wien auf gemeinsame Kosten mit Dr. Wilhelm Pelizaeus unternommenen Grabungen auf dem Friedhof des Alten Reiches bei den Pyramiden von Gîza.* Wien: Holder-Pichler-Tempsky. Akademie der Wissenschaften in Wien. Pilosophisch-Historische Klasse.

KAISER, W. (1957). Zur Inneren Chronologie der Naqadakultur. *Archaeologica Geographica, 61,* 67-77.

KAISER, W. (1985). Zur Suausdehnung der vorgeschichtlichen deltakulturen und frühen entwicklung oberägyptens. *Mitteilungen des Deutchen Archäeologischen Instituts Abteilung Kairo (MDAIK), 41,* 61-87.

KAISER, W., AVILA, R., DREYER, G., JARITZ, H., ROSING, F. W., & SEIDLMAYER, S. (1982). Stadt und Tempel von Elephantine. *Mitteilungen des Deutchen Archaologischen Insitituts, Abteilung Kairo (MDAIK), 38,* 271-345.

KAMRIN, J. (1999). *The Cosmos of Khnumhotep II at Beni Hasan.* New York: Keagan Paul International.

KELSO, J. L., & THORLEY. (1943). The Potter's Technique at Tell Beit Mirsim, Particularly in Stratum A. In W. F. Albright (Ed.), *The Excavations of Tell Beit Mirsim, Vol III (4), The Iron Age* (pp. 86-119). Chicago: Annals of the American School of Oriental Research.

KEMP, B. (1966). Abydos and the Royal Tombs of the First Dynasty. *The Journal of Egyptian Archaeology, 52,* 13-22.

KEMP, B. (1987). Report on the 1986 excavations: chapel 556. In B. J. Kemp, *Amarna Reports IV* (pp. 70-86). London.

KEMP, B. (1995). *Amarna VI.* London.

KEMP, B. (2006). *Ancient Egypt. Anatomy of a Civilisation.* Routledge Press.

KEMP, B., & ROSE, P. (1991). Proportionality in Mind and Space in Ancient Egypt. *Cambridge Archaeological Journal* (1), 103-29.

KENYON, K. M., & HOLLAND, T. A. (1983). *Excavations at Jericho. Vol Five. The Pottery Phases at the Tell and Other Finds.* Jerusalem: British School of Archaeology in Jerusalem.

KLEMM, R., & KLEMM, D. D. (1993). *Stein und steinbrüche im alten Ägypten.* Berlin: Springer-Verlag.

KNAPP, A. B. (1988). *The History and Culture of Western Asia and Egypt.* The Dorsey Press.

KOCKELMANN, H. (2011). Mammisi (Birth House). *UCLA Encyclopedia of Egyptology,* 1-7.

KÖHLER, E. C. (1992). Problems and Priorities in the study of Pre and Early Dynastic Pottery. In *Cahiers de la Céramique Egyptienne III* (pp. 7-15).

KÖHLER, E. C. (1995). The State of Research on Late Predynastic Egypt: New Evidence for the Development of the Pharaonic State. *Göttinger Miszellen,* 79-92.

KÖHLER, E. C. (1998). *Tell el-Fara'in-Buto III, Die Keramik von der späten Naqada-Kultur bis zum frühen Alten Reich (Schichten III bis VI).* Mainz: Philipp von Zabern, AVDAIK 94.

KREJČÍ, J. (2000). The Origins and Development of the Royal Necropolis at Abusir during the Old Kingdom. In M. Bárta, & J. Krejčí (Eds.), *Abusir and Saqqara in the year 2000.* (pp. 467-84). Praha: ArOr Suppl. IX.

KREJČÍ, J. (2009). *Abusir XI: the architecture of the mastaba of Ptahshepses.* Praha: Czech Institute of Egyptology.

KROEBER, A. L. (1957). *Style and Civilisation.* Ithaca: Cornell University Press.

KROEPER, K. (1989). Palestinian and Ceramic Imports in Pre- and Proto-historic Egypt. In P. de Miroschedjii (Ed.), *L'urbanisation de la Palestine à l'Age du Bronze Ancien. Bilan et perspectives des researches actuelles.* (pp. 407-420). Actes du Colloque d'Emmais (20-24 octobre 1986) Past II. BAR International Series 527 (ii).

KROEPER, K. (1992). Shape + Matrix = Workshop? Ceramic from Minshat Abu Omar. *Cahiers Ceramique d'Egyptienne,* 23-31.

KROEPER, K., & WILDUNG, D. (1994). *Minshat Abu Omar. Ein vor- und fruhgeschichtlicher Friedhof im Niledelta* (Vol. I). Mainz.

KUHRT, A. (1995). *The Ancient Near East, c3000-300 B.C.* (Vol. 1). London and New York: Routlege Press.

LARSEN, H. (1941). Vorbericht uber die schwedischen Grabungen in Abu Ghalib 1936/1937. *Mitteilungen des Deutschen Instituts fur Agyptische Altertumskunde in Kairo (MDAIK), 10,* 1-59.

LAUFER, B. (1917). *The Beginnings of Procelain in China.* Chicago: Anthropological Series, Vol 15. No. 2. Field Museum of Natural History.

LEACH, B. (1945). *A Potter's Book.* London: Faber and Faber Ltd.

LECLANT, J., MATHIEU, B., & PIERRE-CROISIAU, I. (2001). *Les textes de la pyramide de Pépy Ier* (Vols. Tomes 1-2). Le Caire: MIFAO 118/1-2.

LECUYOT, G., & PERRAT, G. (1992). À Propos Des Lieux de Production de Quelques Céramiques Trouvées à Tôd et dans la Vallée des Reines. In *Cahiers de la Céramique Égyptienne 3. Ateliers de Potiers et Productions Céramiques en Égypte* (pp. 173-180). Le Caire: Insituts Français D'Archéologie Orientale du Caire.

LEHNER, M. (2008). *The Complete Pyramids: Solving the Ancient Mysteries* (Vol. 1997 reprint). London: Thames and Hudson.

LEHNER, M. (2010). Chapter 5: Villages and the Old Kingdom. In W. Wendrich (Ed.), *Egyptian Archaeology* (pp. 85-101). Chichester: Wiley-Blackwell.

LEMONNIER, P. (1980). *Les Salines de l'Ouest: Logique technique, logique sociale.* Paris/Lille: Editions de la Maison des science de l'homme, Presses Universitaires de Lille.

LEMONNIER, P. (1989). Towards an Anthropology of Technology. *Man*, 526-527.

LEMONNIER, P. (2002). *Technological Choices. Transformations in material culture since the Neolithic* (first published 1993 ed.). (P. Lemonnier, Ed.) London: Routledge Press.

LEROI-GOURHAN, A. (1943/5). *Evolution et Techniques* (Vol. I: Homme et la Matiere and II: Milieu et Techniques). Paris: Albin Michel.

LÉVI-STRAUSS, C. (1973). *Anthropologie Structurale II.* Paris: Plon.

LEVY, T. E., & ALON, D. (1985). Shiqmim: A Chalcolithic Village and Mortuary Centre in the Northern Negev. *Paléorient*, 11 (1), 71-83.

LEVY, T. E., & HOLL, A. (1988). Les Sociétés Chalcolithiques de la Palestine et l'émergence de chefferies. *Archives Européenes de Sociologie*, 29 (2), 283-316.

LICHTHEIM, M. (1975). *Ancient Egyptian Literature: The Old and Middle Kingdoms* (Vol. I). University of California Press.

LITTO, G. (1976). *South American Folk Pottery.* Watson-Guptill Publications.

LIVINGSTONE-SMITH, A. L. (2000). Processing Clay for Pottery in Northern Cameroon: social and technical requirements. *Archaeometry*, 21-42.

LONGACRE, W. A. (1999). Standardization and Specialisation: What's the link? In J. M. Skibo, & G. M. Feinman (Eds.), *Pottery and People. A Dynamic Interaction* (pp. 44-58). Salt Lake City: University of Utah Press.

LOUD, G. L. (1948). *Megiddo: Seasons of 1935-1939* (Vol. II).

LUCAS, A. (1932). Black and Black-topped pottery. *Annales du Service des Antiquités Egyptiennes.*

LUCAS, A., & HARRIS, J. R. (1962). *Ancient Egyptian Materials and Industries* (4th ed.). London: Edward Arnold Ltd.

LUSTIG-ARECCO, V. (1975). *Technology: Strategies for Survival.* New York: Rhinehart and Winston.

MACE, A. C. (1922). The Egyptian Expedition 1921-1922, part 2. *Supplement to the Bulletin of the Metropolitan Museum of Art.*

MACKENZIE, D., & WAJCMAN, J. (Eds.). (1999). *The Social Shaping of Technology* (2nd edition ed.). Maidenhead, Berkshire: Open University Press.

MACLEAN, R. (1998). Gendered Technologies and Gendered Activities in the Interlacustrine Early Iron Age. In S. Kent (Ed.), *Gender in African Prehistory* (pp. 163-177). Walnut Creek, London, New Delhi: AltaMira Press.

MACZYNSKA, A. (2004). Pottery Tradition at Tell el-Farkha. In S. Hendrickx, R. Friedman, K. M. Cialowicz, & M. Chlodnicki (Eds.), *Egypt at its origins: studies in memory of Barbara Adams* (pp. 421-441). Louvain, Belgium: Peeters Publishers.

MAGRILL, P., & MIDDLETON, A. (1997). A Canaanite Potter's Workshop in Palestine. In I. Freestone, & D. Gaimster (Eds.), *Pottery in the Making, world ceramic traditions* (pp. 68-73). London: Britsh Museum Press.

MAISLER, B., STEKELIS, M., & AVI-YONAH, M. (1952). The Excavations at Beth Yerah (Khirbet el-Kherak) 1944-1946. *Israel Exploration Society*, 2 (3), 165-173.

MAJCHEREK, G., & EL-SHENNAWI, A. e.-A. (1992). Research on Amphorae Production on the norrthwestern coast of Egypt. In P. Ballet (Ed.), *Cahiers de la Céramique Égyptienne 3. Ateliers de Potiers et Productions Céramiques en Égypte* (pp. 129-136). Le Caire: Insituts Français D'Archéologie Orientale du Caire.

MALLORY-GREENOUGH, L. M., GREENOUGH, J. D., & OWEN, J. V. (1999). The Stone Source of Predynastic Basalt Vessels: Mineralogical Evidence for Quarries in Northern Egypt. *Journal of Archaeological Science* (26), 1261-1272.

MALLORY-GREENOUGH, L. M., GREENOUGH, J. D., & OWEN, J. V. (2000). The Origin and Use of Basalt in Old Kingdom Funerary Temples. *Geoarchaeology: An International Journal*, 15 (4), 315-330.

MARK, S. (1997). *From Egypt to Mesopotamia: a Study of Predynastic Trade Routes.* Texas: AltaMira Press.

MARTIN-PARDEY, E. (1984). Scheingaben. *Lexicon Agypten*, V, 560-3.

MATTHEWS, R. (2003). *The Archaeology of Mesopotamia. Theories and Approaches.* London and New York: Routledge Press.

MAUSS, M. (1936). Les Techniques du Corps. *Journal de Psychologie*, 365-86.

MAYR, O. (1976). The Science-Technology Relation as a Historiographic Problem. *Technology and Culture*, 17, 663-72.

MCADAM, E., & MYNORS, H. S. (1988). Tell Rubeidheh: potteryb from the Uruk mound. In R. G. Killick (Ed.), *Tell Rubeidheh: an Uruk village in the Jebel Hamrin* (pp. 39-76). Warminister: Aris &Phillips/British School of Archaeology in Iraq and the Directorate of Antiquities.

MCNICOLL, A., SMITH, R., & HENNESSY, B. (1982). *Pella in Jordan I: An Interim Report on the Joint Expedition of Sydney and the College of Wooster Excavations at Pella 1979-1981.* Canberra: The Sydney National Gallery.

MIDANT-REYNES. (2000). The Naqada Period (c4000-3200 B.C.). In I. Shaw (Ed.), *The Oxford history of Ancient Egypt* (pp. 44-60). Oxford: Oxford University Press.

MILLER, H. M.-L. (2009). *Archaeological Approaches to Technology*. Walnut Creek: Left Coast Press Inc.

MILWARD JONES, A. (1991). Section D, Pottery. In A. el-Khouli, & G. T. Martin (Ed.), *Meidum* (pp. 43-45). The Australian Centre for Egyptology: Reports 3.

MOOREY, P. R. (1987). On tracking cultural Transfers in Prehistory: the case of Egypt and Lower Mesopotamia in the fourth millennium B.C. In M. Rowlands, M. Larsen, & K. Kristiansen (Eds.), *Centre and Periphery in the Ancient World* (pp. 36-46). Cambridge: Cambridge University Press.

MORGAN, L. H. (1877). *Ancient Society*. New York: World.

MURDOCK, G. P., & PROVOST, C. (1973). Factors in the Division of Labor by Sex: A Cross-Cultural Analysis. *Ethnology, Vol. 12* (2), 203-225.

MYŚLIWIEC, K., & POŁUDNIKIEWICZ, A. (2003). A Center of Ceramic Production in Ptolemaic Athribis. In C. A. Redmont, & C. A. Keller (Eds.), *Egyptian Pottery. Proceedings of the 1990 Pottery Symposium at the University of California, Berkeley* (pp. 128-152). Berkeley: Contributions of the the University of California Archaeology Research Facility, no. 58.

NEWBERRY, P. E. (1893). *Beni Hasan part 1*. London: Egypt Exploration Fund.

NEWBERRY, P. E; FRASER. (1894). *Beni Hasan Part II*. London: Egypt Exploration Fund.

NEWBERRY, P. E., & GRIFFITH, F. L. (1895). *El Bersheh, Part I (The Tomb of Tehuti-hetep)*. London.

NIBBI, A. (1987). *Ancient Egyptian Pot Bellows and the Oxhide Ingot*. Oxford: D. E. Publications.

NICHOLAS, I. M. (1987). The function of bevelled-rim bowls: a case study at the TUV mound. *Paléorient, 13* (2), 61-72.

NICHOLSON, P. T. (1989). Ceramic Technology in Upper Egypt: a study of pottery firing. *World Archaeology, 21* (1), 71-86.

NICHOLSON, P. T. (1992). The Pottery Workshop at El-Amarna. *Cahiers Ceramique d'Egyptienne, 3*, 61-70.

NICHOLSON, P. T. (1993). The Firing of Pottery. In D. Arnold, & J. Bourriau, *An Introduction to Ancient Egyptian Pottery* (pp. 103-120). Mainz am Rhein: Verlag Philipp von Zabern.

NICHOLSON, P. T. (1995a). The Potters of Deir Mawas, an Ethnoarchaeological Study. In B. Kemp (Ed.), *Amarna Reports Vol. VI*, (pp. 279-299).

NICHOLSON, P. T. (1995b). Kiln Excavations at P47.20 (House of Ramose Complex). In B. Kemp (Ed.), *Amarna Reports VI* (pp. 226-238). London.

NICHOLSON, P. T. (2002). Deir Mawas and Deir el Gharbi: Two Ceramic Traditions. In W. Z. Wendrich, & G. van der Kooij (Eds.), *Moving Matters: Ethnoarchaeology in the Near East*. (pp. 139-146). Leiden.

NICHOLSON, P. T. (2009). Pottery Production. *Encyclopedia of Egyptology*, 1-8.

NICHOLSON, P. T. (2010). Kilns and Firing Structures. *UCLA Encyclopedia of Egyptology*, 1-10.

NICHOLSON, P. T., & DOHERTY, S. K. (forthcoming). Arts and Crafts: Artistic Representations as a Guide to Craft Technique. In *Vienna 2: Egyptian Ceramics in the 21st Century*.

NICHOLSON, P. T., & PATTERSON, H. L. (1989). Pottery making in Upper Egypt: an ethnoarchaeological study. *World Archaeology, 17* (2), 222-39.

NICKLIN, K. (1979). The Location of Pottery Manufacture. *Man, 14* (3), 436-458.

NISSEN, H. J. (1970). Grabung in den Quadraten K/L XII in Uruk-Warka. *Baghdader Mitteilungen*, 137.

NISSEN, H. J. (1988). *The Early History of the Ancient Near East 9000-2000 B.C.* Chicago: University of Chicago Press.

O'CONNOR, D. (2011). *Abydos: Egypt's First Pharaohs and the Cult of Osiris*. London: Thames and Hudson.

ODLER, M. (in press). The palaeographic and iconographic sources for the potter's wheel of the Old Kingdom. In T. Rzeuska, & A. Wodzinska (Eds.), *Old Kingdom Pottery Workshop: Chapter 2*. Warsaw.

OGDEN, J. (2000). Metals. In P. T. Nicholson, & I. Shaw (Eds.), *Ancient Egyptian Materials and Technology* (pp. 148-176). Cambridge: Cambridge University Press.

OP DE BEECK, L. (2004). Possibilities and Restrictions for the Use of Maidum-Bowls as Chronological Indicators. *Cahiers de la ceramique Égyptienne, 7*, 239-275.

OREN, E. D. (1987). The "Ways of Horus" in North Sinai. In A. F. Rainey, *Egypt, Israel and Sinai: Archaeological and Historical Relationships in the Biblical Period* (pp. 69-119). Tell Aviv.

OREN, E. D., & YEKUTIELI, Y. (1992). Taur Ikhbeineh: Earliest Evidence for Egyptian Interconnections. In E. C. van der Brink (Ed.), *The Nile Delta in Transistion, 4th-3rd Millennium B.C.* (pp. 361-384). Tel Aviv: Israel Exploration Society.

PANTALACCI, L. (1998). La documentation épistolaire du palais des gouverneurs à Balat-Ayn Asil. *BIFAO, 98*, 303-15.

PAPAZIAN, H. (2005). *Domain of the Pharaoh: The Structure and Components of the Economy of the Old Kingdom*. Chicago: Unpublished PhD Thesis, University of Chicago.

PARKINSON, R. (1999). *The Tale of Sinuhe and other Ancient Egyptian Poems 1940-1640 B.C.* Oxford: World's Classics.

PATRIK, L. (1985). Is there an archaeological record? In M. Schiffer, *Advances in Archaeological Method and Theory* (pp. 27-62). Academic Press.

PAZ, S. (2006). Chapter 3: Area SA: The Stekelis-Avi-Yonah Excavations (Circles Building) 1945-1946. In R. Greenberg, E. Eisenberg, S. Paz, & Y. Paz (Eds.),

Bet Yerah: The Early Bronze Age Mound. Volume I. Excavation Reports 1933-1986 (pp. 53-104). Jerusalem: The Israel Antiquities Authority.

PEACOCK, D. P. (1977). *Pottery and early commerce: characterization and trade in Roman and later ceramics.* London & New York: Academic Press.

PEACOCK, D. P. (1981). *Pottery in the Roman world: an ethnoarchaeological approach.* London: Longman.

PEET, T. E., & WOOLLEY, L. (1923). *City of Akhenaton Part I.* London: The Egypt Exploration Society, 38th memoir.

PELEGRIN, J. (1990). Prehistoric Lithic Technology: Some Aspects of Research. *Archaeological review from Cambridge, 9* (1), 116-125.

PELEGRIN, J., KARLIN, C., & BODU, P. (1988). Chaîne Opératoires: un outil pour le préhistorien. In J. Tixier (Ed.), *Technologie Préhistorique* (pp. 55-62). Paris: CNRS.

PELTA, R. (1996). A Potter's Wheel from Tel Dalit. In R. Gophna (Ed.), *Excavations at Tel Dalit* (pp. 171-185). Tel Aviv: Ramot Publications.

PENDLEBURY, J. D. (1951). *City of Akhenaten, Part III* (Vol. 2). London: Egypt Exploration Society, 44th Memoir.

PEREGRINE, P. (1991). Some political aspects of craft specialization. *World Archaeology* (1), 1-11.

PERROT, J., & LADIRAY, D. (1980). *Tombes à Ossuaires de la Région Côtière Palestinienne au IVe millénaire avant l'Ere Chrétienne,.* Paris: Association Paléorient.

PETRIE, W. M. (1891). Egyptian Exploration. The Oldest Pyramid and Temple. *The Academy, 989,* 376.

PETRIE, W. M. (1892). *Medum.* London: David Nutt, Strand.

PETRIE, W. M. (1894). *Tell el Amarna.* London.

PETRIE, W. M. (1888). *Tanis, Part II: Nebesheh (Am) and Defenneh (Tahpanhes).* London: EEF.

PETRIE, W. M. (1911). The pottery Kilns at Memphis. In E. B. Knobel, *Historical Studies* (pp. 34-37). London.

PETRIE, W. M. (1902). *Abydos I.* London.

PETRIE, W. M. (1921). *Corpus of Prehistoric Pottery and Pallettes.* London: British School Archaeology in Egypt.

PETRIE, W. F. (1925). *Ancient Egyptians.* London: Williams and Norgate.

PETRIE, W. M. (1977). *The Funeral Furniture of Egypt with stone and metal vases.* Warminster: Aris and Phillips Ltd.

PETRIE, W. M., & MACE, A. C. (1901). *Diospolis Parva. The Cemeteries of Abadiyeh and Hu 1898-1899.* London.

PETRIE, W. M., & QUIBELL, J. (1896). *Naqada and Ballas.* London: BSAE I.

PETRIE, W. M., MACKAY, E., & WAINWRIGHT, G. (1910). *Meydum and Memphis (III).* London: British School of Archaeology in Egypt and Egyptian Research Account. UCL & Bernard Quaritch.

PIERRE, I. (1997). Les signes relatifs à l'homme dans les Textes des Pyramides, Quelques particularités et graphies inhabituelles, jeux graphiques et fautes d'ortographie, in: C. Berger – B. Mathieu (. In C. Berger, & B. Mathieu (Eds.), *Études sur l'Ancien Empire et la nécropole de Saqqâra dédiées à Jean-Philippe Lauer.* (Vol. 2, pp. 355-360). Orientalia Mospeliensia IX, Montpellier.

PLOG, F. (1974). *The Study of Prehistoric Change.* New York: Academic Press.

PLOG, S. (1980). *Sylistic Variation in Prehistoric Ceramics: Design Analysis in the Americas.* Cambridge: Cambridge University Press.

POLLOCK, S. (1999). *Ancient Mesopotamia: The Eden that Never Was.* Cambridge: Cambridge University Press.

PORAT, N. (1992). Presence of Egyptians in Palestine during EB I. In E. C. van den Brink (Ed.), *The Nile Delta in Transition 4th-3rd millennium B.C.* (pp. 433-440). Tel Aviv: Israel Exploration Society.

PORTER, B., & MOSS, R. (1972). *Topographical bibliography of Ancient Egyptian Hieroglyphic Texts, Reliefs and Paintings. II Theban Temples, 2nd Editiion.* Oxford: Griffith Institute.

PORTER, B; MOSS, R; MALEK, J (2003): *Topgraphical Bibliograpy pf Ancient Egyptian Hiéroglyphic Texts, Reliefs and Paintings, III part 2, Memphis: Saqqâra to Dahshûr,*Oxford, Griffith institute

PORTER, B., & MOSS, R. (2004). *Topographical Bibliography of Ancient Egyptian Hieroglyphic Texts, Reliefs and Paintings, IV Lower and Middle Egypt.* Oxford, Chippenham: Griffith Insitute, Oxford (1934 reprint).

POSENER-KRIÉGER, P. (1976). *Les archives du temple funéraire de Néferirkarê-Kakaï (les papyrus d`Abousir), Traduction et commentaire I-II* (Vol. 65). Le Caire: Intitute Français d'Archéoligie Orientale.

POSENER-KRIÉGER, P.(2004) *I papiri di Gebelein – Scavi G. Farina 1935,* Studi del Museo egizio di Torino: Gebelein, vol. 1, Torino.

POSENER-KRIÉGER, P., & DE CENIVAL, J. L. (1968). *Hieratic Papyri in the British Museum. Fifth Series. The Abusir Papyri.* London: The Trustees of the British Museum.

POSENER-KRIÉGER, P., VERNER, M., & VYMAZALOVÁ, H. (2006). *Abusir X. The Pyramid Complex of Raneferef Papyrus Archive.* Prague: Czech Insitute of Egyptology.

POWELL, C. (1995). The Nature and Use of Ancient Egyptian Potter's Wheels. In *Amarna Reports VI* (pp. 309-335). London.

PRITCHARD, A., & VAN DER LEEUW, S. E. (1984). Introduction: The Many Dimensions of Pottery. In A. Pritchard, & S. E. van der Leeuw (Eds.), *The Many Dimensions of Pottery: Ceramics in Archaeology and Anthropology* (pp. 3-23). Amsterdam: University of Amsterdam.

QUIBELL, J. E. (1900). *Hierakonpolis I.* London: Bernard Quaritch.

QUIBELL, J. E. (1908). *Excavations at Sakkara 1906-1907.* Cairo.

QUIBELL, J. E., & GREEN, F. W. (1902). *Hierakonpolis* (Vol. II). London: ERA 5.

QUIBELL, J. E., & HAYTER, A. G. (1927). *Teti Pyramid, North Side.* Cairo.

RADO, P. (1969). *An Introduction to the Technology of Pottery.* Oxford: Paragon Press Ltd.

REFORD, D. B. (1981). Interim Report on the excavations at East Karnak 1977-1978. *Journal of the American Research Center in Egypt, 18,* 11-41.

REISNER, G. A. (1913). A Family of Builders of the Sixth Dynasty, about 2600 B.C. *Bulletin of the Museum of Fine Arts, Boston, 11* (66), 53-66.

REISNER, G. A. (1923). *Excavations at Kerma.* Cambridge, Mass: Peabody Museum of Harvard University. Harvard African studies vol. V-VI.

REISNER, G. A. (1931). *Mycerinus: The Temples of the Third Pyramid at Giza.* Cambridge, MA: Harvard University Press.

REISNER, G. A. (1934). The History of the Egyptian Mastaba. In *Mélanges Maspero I: Orient Ancien. Mémoires de l'Institut français d'archéologie orientale 66, part 2 (1935–1938),* (pp. 579-584).

REISNER, G. A., & STEVENSON SMITH, W. (1955). *A History of the Giza Necropolis II: The Tomb of Hetep-heres The Mother of Cheops.* Cambridge MA: Harvard University Press.

REYMOND, E. A. (1969). *The Mythical Origin of the Egyptian Temple.* New York: Barnes and Noble.

RICE, P. M. (1981). Evolution of a Specialised Pottery Production: A trial model. *Current Anthropology, 22* (3), 219-240.

RICE, P. M. (1987). *Pottery Analysis; a sourcebook.* Chicago: University of Chicago Press.

RICE, P. M. (1991). Women and Prehistoric Pottery Production. In D. Walde, & N. D. Willows (Eds.), The Archaeology of Gender (pp. 436-443). Calgary: Chacmool. The Archaeological Association of the University of Calgary.

RICE, P. M. (1996). Recent Ceramic Analysis: function, style and origins. *Journal of Archaeological Research, 4,* 113-163.

RICHTER, M. N. (1982). *Technology and Social Complexity.* Albany: State University of New York Press.

RIETH, A. (1960). *5000 Jahre Töpfercheibe.* Konstanz.

RIZKANA, I., & SEEHER, J. (1987). *Maadi I: The pottery of the Predynastic Settlement.* Mainz am Rhein: Verlag Philipp von Zabern.

RIZKANA, I., & SEEHER, J. (1988). *Ma'adi II: The Lithics. Industries of the Predynastic Settlement.* Mainz am Rien: Archaeologische Veroffentlichungen 65. Phillip von Zabern.

ROAF, M. (1990). *Cultural Atlas of Mesopotamia and the Ancient Near East.* Oxford: Equinox Ltd.

ROBINS, G. (1997). *The Art of Ancient Egypt.* London: The Trustees of the British Museum, British Museum Press.

ROCATTI, A. (2006). A chi servivano i Papiri di Gebelein? In *L'ufficio e il documento. I luoghi, i modi, gli strumenti dell'amministrazione in Egitto e nel Vicino Oriente antico,* (pp. 87-91). Milano: Mora, C; Piacentini, P.

ROSE, P. J. (1989). Report on the 1987 Excavations: The Evidence for Pottery Making at Q48.4. In B. Kemp (Ed.), *Amarna Reports Vol. V* (pp. 82-95). London: Egypt Exploration Society.

ROSEN, A. M. (1986). *Cities of clay: the geoarcheology of tells.* Chicago, London: University of Chicago Press.

ROTH, H. L. (1951). *Ancient Egyptian and Greek Looms.* Halifax: Bankfield Museum Notes.

ROUX, V (1990) The Psychological Analysis of Technical Activities: a Contribution to the Study of Craft Specialisation. *Archaeological Review from Cambridge* 9, 142-153.

ROUX, V. (2003). A Dynamic Systems Framework for Studying Technological Change: Application to the Emergence of the Potter's Wheel in the Southern Levant. *Journal of Archaeological Method and Theory, 10* (1), 1-30.

ROUX, V (2008) Evolutionary Trajectories of Technological Traits and Cultural Transmission: A Qualitative Approach to the Emergence and Disappearance of the Ceramic Wheel-fashioning Technique in the Southern Levant. In: M. Stark, B. Bowser and L. Horne (Eds.) *Cultural Transmission and Material Culture. Breaking Down Boundaries.* Tuscon: Arizona University Press, 82-104.

ROUX, V. (2009). Wheel Fashioned Ceramic Production during the Third Millenium B.C.E in the Southern Levant: a Perspective from Tell Yarmuth. In S. A. Rosen, & V. Roux (Eds.), *Tecnhiques and People: Anthropological Perspectives in the Archaeology of Proto-Historic and Early Historic Periods in the Southern Lecvant* (pp. 195-212). Paris: De Boccard.

ROUX, V., & CORBETTA, D. (1989). *The Potter's Wheel. Craft Specialization and Technical Competence.* New Delhi, Oxford: IBH Publishing.

ROUX, V., & COURTY, M. A. (1997). Les bols élaborés au tour d'Abu Hamid: Rupture Technique au 4e millénaire J.-C. dans le Sud Levant. *Paléorient,* 25-43.

ROUX, V., & COURTY, M.-A. (1998). Identification of Wheel-fashioning methods: Technological Analysis of 4th-3rd millennium B.C. Oriental Ceramics. *Journal of Archaeological Science, 25,* 747-763.

ROUX, V., & COURTY, M.-A. (2005). Identifying Social Entities at a Macro-regional level: Chalcolithic Ceramics of South Levant as a Case Study. In A. Livingstone Smith, D. Bosquet, & R. Martineau (Eds.), *Pottery Manufacturing Processes: Reconstitution and Interpretation* (pp. 201-214). Liege: Acts of the XIVth UISPP Congress, University of Liege, Belgium, 2-8 Sept 2001, BAR International Series 1349.

ROUX, V., & DE MIROSCHEDJI, P. (2009). Revisiting the History of the Potter's Wheel in the Southern Levant. *Levant, 41* (2), 115-173.

Rowan, Y. M., & Golden, J. (2009). The Chalcolithic Period of the Southern Levant: A Synthetic Review. *Journal of World Prehistory, 22*, 1-92.

Rowe, A. (1931). Excavations of the Eckley B. Coxe, Jr., Expedition at Meydum, Egypt. *The Museum Journal University of Pennsylvania, XXII.*

Ruscoe, W. (1963). *A Manual for the Potter.* London: Alec Tiranti Ltd.

Rye, O. (1981). *Pottery Technology: Principles and Reconstruction.* Washington D. C.: Taraxacum Press.

Rzeuska, T. (2006a). Saqqara II: Pottery of the Late Old Kingdom. Funerary Pottery and Burial Customs. Varsovie: Polish-Egyptian Archaeology Mission. Editions Neriton.

Rzeuska, T. (2006b). Funerary customs and rites on the Old Kingdom necropolis in West Saqqara. In M. Bárta, F. Coppens, & J. Krejčí (Eds.), *Abu Sir and Saqqara in the year 2005* (pp. 353-377). Prague: Czech Institute of Egyptology.

Rzeuska, T., & Kuraszkiewicz, K. O. (2011). An Offering of a Beer Jar or a Beer Jar as an Offering? The Case of a Late Old Kingdom Beer Jar with an Inscription from West Saqqara. In D. Aston, B. Bader, I. Gallorini, P. Nicholson, & S. Buckingham (Eds.), *Under the Potter's Tree. Studies Presented to Janine Bouriau on the Occasion of her 70th Birthday* (pp. 829-842). Leuven, Paris, Walpole MA: Uitgeveriz Peeters en Departement Oosterse Studies.

Sahlins, M. (1972). *Stone Age Economies.* New York: Aldine-Atherton.

Sanders, W. T., & Price, B. J. (1968). *Mesoamerica: The Evolution of Civilization.* New York: Random House.

Säve-Söderburg, T. (1963). Archaeological Investigations between Faras and Gemai, November 1962-March 1963. *Kush, 2*, 47-99.

Sethe, K. (1910). *Die altägyptischen Pyramidentexte. Zweiter Band. Text, zweite Hälfte, Spruch 469-714 (Pyr. 906-2217).* Leipzig.

Schiffer, M. (1972). Archaeological Context and Systematic Context. *American Antiquity* (37), 156-65.

Schiffer, M. B., & Skibo, J. M. (1987). Theory and Experiment in the Study of Technological Change. *Current Anthropology, 28* (5), 595-622.

Seidlmayer, S. (1988). Funrärer Aufwand und soziale Ungleichheit. Eine methodische Anmerkung zum Problem der Rekonstruktion der gesellschsftlichen Gliederung aus Friedhofsfunden. *Gottinger Miszellen, 104*, 25-51.

Seidlmayer, S. (1990). *Gräberfelder aus dem Übergang vom Alten zum Mittleren Reich.* Heidelberg.

Seidlmayer, S. (2000). The First Intermediate Period. In I. Shaw, *The Oxford History of Ancient Egypt* (pp. 118-147). Oxford: Oxford University Press.

Senussi, A. (2006). The Cemetery's Potter. In E. Czerny, I. Hein, H. Hunger, D. Melman, & Schwab (Eds.), *Timelines: Studies in Honour of Manfred Bietak* (pp. 329-30). Leuven: Peeters.

Serpico, M. (2004). Salvaging the Past: Analysis of First Dynasty Jar Contents in the Petrie Museum of Egyptian Archaeology. In S. Hendrickx, R. F. Friedman, K. M. Cialowicz, & M. Chlodnicki (Eds.), *Egypt at its Origins. Studies in Memory of Barbara Adams. Proceedings of the International Conference "Origins of the State, Predynastic and Early Dynastic Egypt" Krakow 28th August 2002* (p. 2004). Leuven, Paris, Dudley M.A.: Uit geverij Peeters en departement oostere studies.

Service, E. R. (1962). *Primitive Social Organization.* New York: Random House.

Shaw, I (2000) *The Oxford History of Ancient Egypt.* Oxford, OUP.

Shaw, I. (2004). Identity and Occupation. How did individuals define themselves and their work in the Egyptian New Kingdom? In J. Bourriau, & J. Phillips (Eds.), *Invention and Innovation. The Social Context of Technological Change 2: Egypt, the Aegean, and the Near East. 1650-1150 B.C.E* (pp. 12-24). London, Oxford: Oxbow Books.

Shaw, I. (2011). Seeking the Ramesside Royal Harem: New Fieldwork at Medinet el-Gurob. In M. Collier, & S. Snape (Eds.), *Ramesside Studies in Honour of Ken Kitchen* (pp. 453-63). Bolton: Rutherford Press.

Shennan, S. (1982). Ideology, Change and the European Early Bronze Age. In I. Hodder (Ed.), *Symbolic and Structural Archaeology* (pp. 155-62). Cambridge: Cambridge University Press.

Shepard, A. (1968). *Ceramics for the Archaeologist.* Washington: Carnegie Institution (Publication 609).

Shirai, Y. (2005). Royal Funerary Cults During the Old Kingdom. In K. Piquette, & S. Love (Eds.), *Current Research in Egyptology 2003* (pp. 149--162). Oxford: Oxbow Books.

Simpson, S. J. (1997a). Early Urban Ceramic Industries in Mesopotamia. In D. Gaimster, & F. I (Eds.), *Pottery in the Making: Early Ceramic Traditions* (pp. 50-55). London: British Museum Press.

Simpson, S. J. (1997b). Prehistoric Settlements in Mesopotamia. In D. Gaimster, & I. Freestone (Eds.), *Pottery in the Making. World Ceramic Traditions* (pp. 38-43). London: British Museum Press.

Simpson, W. K., & O'Connor, D. B. (2003). *Slab Stelae of the Giza Necropolis.* New Haven and Philadelphia: Pennsylvania-Yale Expedition to Egypt.

Singer, C., Holymyard, E. J., & Hall, A. R. (1954). *A History of Technology from Early Times to Fall of Ancient Empires* (Vol. I). Oxford: Oxford Clarendon Press.

Skibo, J. M. (1992). *Pottery Function: A Use Alteration Perspective.* London: Plenum.

Smythe, J. (2005). Moments in Mud. *Nekhen News, 17*, 21-3.

SOUKIASSIAN, G., WUTTMANN, M., PANTALACCI, L., BALLET, P., & PICON, M. (1990). *Les Ateiliers de Potiers d'Ayn-Asil.* Cairo: FIFAO 34.

SOWADA, K. (1999). Black-Topped Ware in Early Dynastic Contexts. *The Journal of Egyptian Archaeology, 85,* 85-102.

SPENCER, A. J. (1991). *Death in Ancient Egypt.* London: Penguin.

SPENCER, A. J. (1997). Pottery in Predynastic Egypt. In I. Freestone, & D. Gaimster (Eds.), *Pottery in the Making. World Ceramic Traditions* (pp. 44-55). Washington D.C.: Smithsonian Institute Press.

STADELMANN, R. (1983). Die Pyramiden des Snofru in Dashur. Zweiter Bericht uber die Ausgrabungen an der nordlichen Steinpyramide. *Mitteillungen des Deutschen Archaeologischen Instituts. Abteilung Kairo(MDAIK), 39,* 225-241.

STEIN, G. (1994). Economy, ritual and power in Ubaid Mesopotamia. In G. Stein, & M. S. Rothman (Eds.), *Chiefdoms and Early States in the Ancient Near East. The Organizational Dynamics of Complexity* (Vol. Monographs in World Archaeology 18, pp. 35-46). Madison: Prehistory Press.

STEIN, G. J., & BLACKMAN, J. M. (1993). The Organizational Context of Specialized Craft Production in Early Mesopotamian States. *Research in Economic Anthropology* (14), 29-59.

STEINDORFF, G. (1913). *Das Grab de Ti.* Leipzig: 143 Lichtdrucktafeln und 20 Blattern Hinrichs'sche, J. C. Buchhandlung.

STERLING, S. (2004). *Social Complexity in Ancient Egypt: Functional Differentiation as Reflected in the Distribution of Apparently Standardised Ceramics.* University of Washington: Unpublished PhD thesis.

STOCKS, D. A. (1993) Making stone vessels in ancient Mesopotamia and Egypt. *Antiquity* 67 (no. 256 pp. 596-603).

STOCKS, D. A. (2003). *Experiments in Egyptian Archaeology. Stoneworking technology in Ancient Egypt.* London and New York: Routledge Press.

SUCHMAN, L. A. (1987). *Plans and Situated Actions. The Problem of Human-Machine Communication.* Cambridge: Cambridge University Press.

SWAIN, S. (1995). The Use of Model Objects as Predynastic Egyptian Grave Goods: An Ancient Origin for a Dynastic Tradition. In S. Campbell, & A. Green (Eds.), *The Archaeology of Death in the Ancient Near East* (pp. 35-37). Oxford: Oxbow Monograph 51.

SZABADFALVI, J. (1986). *Hungarian Black Pottery.* Budapest: Corvina.

TABOR, D. (1954). Mohs's Hardness Scale-A Physical Interpretation. *Proceedings of the Physics Society, Section B, 67* (3), 249-257.

TAKAMIYA, I. (2004). Development of Specialisation in the Nile Valley During the 4th Millennium B.C. In S. Hendrickx, R. F. Friedman, K. M. Cialowicz, & M. Chlodnicki (Eds.), *Egypt at its Origins. Studies in Memory of Barbara Adams. Proceedings of the International Conference "Origin of the State. Predynastic and Early Dynastic Egypt", Krakow, 28th August - 1st September 2002* (pp. 1027-1039). PEETERS.

TAKAMIYA, I. H., & BABA, M. (2004). Kilns in Square A6: The Other Side of the Story. *Nekhen News, 16,* 19-20.

TAYLOR, J. H. (2001). *Death and the Afterlife in Ancient Egypt.* London: The Trustees of the British Museum.

TEETER, E. (2003). *Ancient Egypt. Treasures from the collection of the Oriental Institute University of Chicago.* Chicago: Oriental Institute Museum Publications no. 23.

TOOLEY, A. M. (1995). *Egyptian Models and Scenes.* Princes Risborough: Shire Publications Ltd.

TORCZYNER, H. (1938). *Lachish: (Tell ed Duweir) 1, The Lachish Letters* (Vol. 1). London: published for the Trustees of the late Sir Henry Wellcome by the Oxford University Press: Wellcome Archaeological Research Expedition to the Near East.

TOSI, M. (1984). The notion of craft specialization and its representation in the archaeological record of early states in the Turanian Basin. In M. Spriggs (Ed.), *Marxist perspectives in archaeology* (pp. 22-52). Cambridge: New Directions in Archaeology, Cambridge University.

TRIGGER, B. (1983). The Rise of Egyptian Civilisation. In B. Trigger, B. Kemp, D. O' Connor, & A. Lloyd (Eds.) *Ancient Egypt: a social history* (pp. 1-70). Cambridge: Cambridge University Press.

TUFNELL, O., MURRAY, M., & DIRINGER, D. (1953). *Lachish: Tell ed Duweir 3, The Iron Age* (Vol. 3). London: Published for the late Sir Henry Wellcome by the Oxford University Press: Wellcome-Marston Archaeological Research Expedition to the Near East.

TUFFNELL, O. (1958). *Lachish IV: The Bronze Age Texts.* Oxford: Oxford University Press. The Trustees of the Late Sir Henry Welcome.

TURNEY-HIGH, H. H. (1949). *General Anthropology.* New York.

USSISHKIN, D. (1980). The Ghassulian Temple at En-Gedi. *Tel Aviv, 7,* 1-44.

VACHALA, B. (2004a). *Abusir VIII: die reliefragmente aus der Mastaba des Ptahschepses in Abusir.* Praha: Excavations of the Czech Institute of Egyptology.

VACHALA, B. (2004b). *Abusir: Die Reliefs aus der Ptahschepses-Mastaba in Abusir Vol III (Excavations of the Czech Institute of Egyptology).* Praha.

VACHALA, B., & FALTINGS, D. (1995). Töpferei und Brauerei im AR- einge Relieffragmente aus der Mastaba des Ptahshepses in Abusir. *Mitteilungen des Deutschen Archäologischen Intituts Abteilung Kairo (MDAIK), 55,* 281-286.

VALBELLE, D. (1985). Les Ouviers de la tombe. Deir el Médineh à l'époque ramesside. *Bulletin d'Egypte, 96.*

VAN BUREN, E. D. (1952). Places of Sacrifice "Opferstatten". *Iraq, 14,* 76-92.

VAN DEN BRINK, E. C. (1989). A Traditional Late Predynastic-Early Dynastic settlement site in the northeastern Nile Delta, Egypt. *Mitteilungen des Deutschen Archäologischen Instituts (MDAIK)* (45), 55-108.

VAN DER KOOIJ, G., & WENDRICH, W. (2002). The Potters of el-Fustat (Cairo) and el-Nazla (Fayoum). In G. van der Kooij, & W. Wendrich (Eds.), *Moving Matters. Ethnoarchaeology of the Near East. Proceedings of the International Seminar held in Cairo. 7-10 December 1998* (pp. 147-158). Leiden: Research School of Asian, African and Amerindian Studies, Universiteit, Leiden.

VAN DER LEEUW, S. E., PAPOUSEK, D. A., & COUDART, A. (1991). Technical Traditions and Unquestioned Assumptions: The Case of Pottery in Michoacan. *Techniques et Culture, 17-18,* 145-173.

VAN DER LEEUW, S. (1976). *Studies in the Technology of Ancient Pottery.* Amsterdam: Organisation for the advancement of Pure Research, Universieit van Amsterdam.

VAN DER LEEUW, S. (2002). Giving the potter a choice, conceptual aspects of pottery techniques. In P. Lemonnier (Ed.), *Technological Choices. Transformations in material culture since the Neolithic* (pp. 238-288). London: Routledge Press.

VAN ELSBERGEN, M. J. (1997). *Fischerei im alten Ägypten. Untersuchungen zu den Fischfangdarstellungen in den Gräbern der 4. bis 6 Dynastie.* Berlin: Achet Verlag,.

VANDIVER, P., & LACOVARA, P. (1985). Outline of Technological Changes in Egyptian Pottery Manufacture. *Bulletin of the Egyptological Seminar, 7,* 53-85.

VARILLE, A., & ROBICHON, C. (1935). Quatre nouveaux temples thebains. *Chronique d'Egypte, 10,* 237-242.

VASILJEVIĆ, V. (2003). Terminology and Interpretation in Studies on Decoration on Private Tombs. *Journal of the Serbian Archaeological Society, 19,* 135-42.

VERCOUTTER, J. (1970). *Mirgissa I.* Paris.

VEREECKEN, S. (2011). An Old Kingdom Bakery at Sheikd Said South: Preliminary Report on the Potter Corpus. In N. Strudwick, & H. Strudwick (Eds.), *Old Kingdom, New Perspectives. Egyptian Art and Archaeology* (pp. 278-285). Oxford and Oakville: Oxbow Books.

VERNER, M. (1992). The Discovery of a Potter's Workshop in the Pyramid Complex of Khentkaus at Abusir. *Cahiers de la Céramique de la Égyptienne 3, 3,* pp. 55-59.

VERNER, M. (1995). *Abusir III: The Pyramid Complex of Khentkaus.* Prague.

VINCENTELLI, M. (2003). *Women Potters. Transforming Traditions.* New Brunswick, New Jersey: Rutgers University Press.

VOGELSANG-EASTWOOD, G. (2000). Textiles. In P. T. Nicholson, & I. Shaw (Eds.), *Ancient Egyptian Materials and Technology* (pp. 268-298). Cambridge: Cambridge University Press.

VON DER WAY, T. (1987). Tell el Fara'in- Buto 2. Bericht. *Mitteilungen des Deutschen Archäologischen Instituts, Abteilung Kairo (MDAIK), 43,* 241-257.

VON DER WAY, T. (1992). Indications of Architecture with Niches at Buto. In R. Friedman, & B. Adams (Eds.), *Followers of Horus: Studies Dedicated to Michael Allen Hoffmann, 1944-1990* (pp. 217-226). Oxford: Oxbow Books, Monograph 20, Egyptian Studies Association Publication no. 2.

VON DER WAY, T. (1997). *Tell el-Fara'în Buto 1, Ergebnisse zum frühen Kontext. Kampagnen der Jahre 1983-1989.* Mainz: Philiip von Zabern.

VYMAZALOVÁ, H. (2011). The Economic Connection Between the Royal Cult in the Pyramid Temples and the Sun Temples in Abusir. In N. Strudwick, & H. Strudwick (Eds.), *Old Kingdom, New Perspectives. Egyptian Art and Archaeology* (pp. 295-303). Oxford: Oxbow.

WALSEM, R. (2005). *Iconography of Old Kingdom Elite Tombs: Analysis & Interpretation, Theoretical and Methodological Approaches.* Leuven: Peeters.

WARD, W. (1982). *Index of Egyptian Administrative and Religious Titles of the Middle Kingdom.* Beirut: American University of Beirut.

WARDEN, L. A. (2010). *The Relationship of Pottery and Economy in Old Kingdom Egypt: A question of state control.* Pennsylvania: Unpublished PhD thesis, University of Pennsylvania.

WENDRICH, W. (1999). *The World According to Basketry. An Ethno-archaeological Interpretation of Basketry Production in Egypt.* Leiden: Research School of Asian, African and Amerindian Studies (CNWS), Universiteit Leiden.

WENDRICH, W. (2006). Body Knowledge Ethnoarchaeological Learning and the Interpretation of Ancient Technology. In *L'apport de l'Égypte à l'histoire des techniques* (pp. 267-275).

WENKE, R. J. (2009). *The Ancient Egyptian State. The Origins of Egyptian Culture (c8000-2000 B.C.).* Cambridge: Campbridge University Press.

WILKINSON, T. A. (2001). *Early Dynastic Egypt* (1999 reprint ed.). London: Routledge.

WILLEMS, H., VEREECKEN, S., KUIJPER, L., VANTHUYNE, B., MARINOVA, E., LINSEELE, V., et al. (2009). An Industrial Site at al-Shaykh Sa'id/Wadi Zabayda. *Egypt and the Levant, 19,* 293-331.

WINNER, L. (1999). Do Artifacts Have Politics? In M. D, & J. Wajcman (Eds.), *The Social Shaping of Technology* (pp. 28-40). Maidenhead, Berkshire: The Open University Press.

WODZIŃSKA, A. (2006). White carinated bowls (CD7) from the Giza Plateau Mapping Project: Tentative typology, use and origin. In M. Bárta, F. Coppens, & J. Krejčí (Eds.), *Abusir and Saqqara in the Year 2005, Proceedings of the International Symposium, Prague 27–30 June 2005*, (pp. 405-429). Prague.

WODZIŃSKA, A. (2007) Preliminary Ceramic Report. In: M. Lehner and W. Wetterstrom (Eds.) *Project History, Survey, Ceramics and the Main Street and Gallery III.4 Operations. Giza Reports 1.* (pp. 275-315) Boston: Ancient Egypt Research Associates, Inc.

WODZIŃSKA, A. (2009a). Work Organization in the Old Kingdom pottery workshop: The case of the Giza Plateau Mapping Project Pottery. In T. I. Rzeuska, & A. Wodzińska (Eds.), *Studies in Old Kingdom Pottery* (pp. 225-240). Warsaw: Wydawnictwo Neriton.

WODZIŃSKA, A. (2009b). Potmarks from Early Dynastic Buto and Old Kingdom Giza: Their occurence and economic significance. *British Museum Studies in Ancient Egypt and Sudan, 13*, 239-61.

WODZIŃSKA, A. (2009c). *A Manual of Egyptian Pottery: Naqada III - Middle Kingdom (Vol. II).* London: Ancient Egypt Research Associates Oxbow books.

WOOLLEY, L. C. (1955). *Ur Excavations* (Vol. IV). Oxford: The Trustees of the British Museum and Museum of the University of Pennsylvania to Mesopotamia.

WOOD, B. G. (1990). *The Sociology of Pottery in Ancient Palestine. The Ceramic Industry and the Diffusion of Ceramic Style in the Bronze and Iron Ages.* Sheffield: Journal for the Study of Old Testament Studies Supplement Series 103. JSOT/ASOR Monographs 4, JSOT Press.

WÜNSCHE, R. (1977). *Studien zur Äginetischen Keramik der Frühen und Mittleren Bronzeezeit.* Berlin: Deutscher Kunstverlag.

XANTHOUDIDES, S. (1927). Some Minoan potter's-wheel discs. In S. Casson (Ed.), *Essays in Aegean Archaeology* (pp. 111-128). Oxford: Clarendon Press.

YADIN, Y. (1958). *Hazor 1, an account of the first season of excavations, 1955.* Jerusalem: Magnes Press, at the Hebrew University.

YADIN, Y. (1960). *Hazor 2, an account of the second season of excavations, 1956.* Jerusalem: Magnes Press, at the Hebrew University.

YEKUTIELI, Y. (2004). The Desert, the Sown, and the Egyptian Colony. *Ägypten und Levante, XIV*, 163-171.

Appendices

The Origins and Use of the Potter's Wheel in Ancient Egypt

Site	Period	Potter's Wheel	Kiln (type)	Other Details	Reference
Mahasna	Pre-Dyn	X	Screen/Pot Kiln	Large vessel supported by firedogs, surrounded at one time by wall	(Garstang, 1902)
Hierakonpolis	Pre-Dyn (c.3650 BCE)	X	Screen /Pot Kiln	Locality 29. Shallow pit-updraught kiln. Large vessel supported by firedogs, surrounded by a low wall. Top not covered	(Hoffman, 1982, p. 12)
Hierakonpolis	Pre-Dyn (c.3200-3100 BCE)	X	Screen/Pot Kiln	Locality IIC, Kiln B1. Shallow pit-updraught kiln. Tamarix and Acacia logs used as fuel	(Harlan, 1982)
Buhen (Northern Town)	OK 4th dyn	X	3 Tall cylindrical circular kilns/copper furnaces	Separate firing chamber by a grid of bricks resting on square support of brickwork. Kilns located to SE. Piles of malachite nearby. Possibly copper furnaces.	(Adams, 1977, pp. 172-3 Emery, 1963, p. 117; Nicholson, 1995)
Abusir (pyr complex of Khentkaus)	OK (4th-5th dyn)	Burnt clay wheel head originally laid on a slab of wood	Circular Kiln/ Conical shape	Wheel in secondary position. Part of the mortuary temple during time of Unas. Workshop surrounded by fence of reed mats. Storeroom. Circular Pit. Kiln to southern opposite end of workshop, built on floor of corridor.	(Verner, 1995, pp. 55-59)
Dakhla Oasis	OK	X	6 Circular Kilns, but poss. 10	Type 1 similar to copper furnaces at Buhen. Type 2 circular or horseshoe shaped with a draught tunnel running from stoke hole. Vessels may have been supported on kiln dogs, more sophisticated version of Hierakonpolis-style. Site 33/390-I9-3 and 33/390-K9-I. Located to the southeast of a small settlement.	(Hope, 1995)
Elephantine	OK (mid 4th-5th dyn)	X	2 Circular kiln	Row of vertical bricks as lowermost course and both open to the north to take advantage of prevailing winds.	(Kaiser, Avila, Dreyer, Jaritz, Rosing, & Seidlmayer, 1982)
Ain Asil (Dakhla Oasis)	OK/FIP-G/R	Poss. 2 pivots, but likely to be for door	25 Circular/horseshoe shape Kilns divided between 2 workshops	Various kilns and associated workshop remains to southwest of main town, kilns belong to 4 phases of use, most open to the south.	(Soukiassian, Wuttmann, Pantalacci, Ballet, & Picon, 1990, pp. 5-9)
Abu Ghalib	MK	X	Circular	Isolated in square, open space, SW of habitation	(Larsen, 1941, p. 11)
Dashur	MK	X	4 Roughly circular	Best preserved Kiln I dimensions 2m E-W by 1.6m N-S. To the northern side is a trench 3m long and 1.2m wide, poss. a draught tunnel to use wind to increase the through-draft. Details of flooring such as T-shaped piece of vaulting found with cross-arm running E-W.	(Stadelmann, 1983, p. 288)
Mirgissa (Sudan)	MK	X	Square Oven, possibly for prep of bread moulds	Opening to hearth is rectangular. Open basin paved with mud in front of kiln	Finnish Egyptological Society, visit 1965; (Holthoer, 1977, p. 16)
Mirgissa (Sudan)	MK	X	Circular	Brick walled pit and fire hole to SE, 2.5m (dia) x 1.0m (depth)	(Vercoutter, 1970, p. fig. 3)
Nag Baba (Sudan)	MK	Poss. pivot/wheel with black lubrication	Screen	Workshop, drying bins, pebbles. Kiln measured 2m x 2m x 1m high.	(Säve-Söderburg, 1963, p. 58)
El-Lahun	MK	X	Circular Pot Kiln		(Mace, 1922, p. fig 15)
Qurnet Mura'i (Amenhotep III's workers village), Thebes	18th dynasty	X	2 large circular	Diameter of c2m, and the stoke hole has a screen towards the south	(Porter & Moss, 1972, p. 457); (Varille & Robichon, 1935)
El-Amarna	18th	X	Square 2 passages at ground, vent holes	Roof of vent holes had 2 diagonally placed bricks. Contained large quantities of charcoal and had a white cobbled pebble floor of quartz	(Petrie, 1894, p. 26)

Appendix I: Selected List of Kilns, Potter's Wheels and Workshops

APPENDIX I: SELECTED LIST OF KILNS, POTTER'S WHEELS AND WORKSHOPS

Site	Period	Potter's Wheel	Kiln (type)	Other Details	Reference
El-Amarna (1986 excavation)	18th	X	Box oven with load of clay bread moulds	Two types of bread mould, conical and chalice. 30 cones stacked in 3 rows of 10. Firing structure rather than just a kiln, as also used to bake bread.	(Frankfort & Pendlebury, 1933) (Kemp, 1987, pp. 73-9);
El-Amarna	18th	X	2 pot kilns	Situated in corner of estate, possibly associated with a kitchen (U.33.9)	(Frankfort & Pendlebury, 1933, p. 74)
El-Amarna North Suburb	18th	Socket and Pivot of granodiorite	X	Associated with the largest house in the Northern Suburb of Amarna T36.11. Ashmolean Museum no.T1929.417	(Hope C., 1981; Powell, 1995, p. 316)
El-Amarna	18th	X	3 pot kilns	Row along S wall of magazines south of the temple	(Pendlebury, 1951, p. 31)
El-Amarna P47.20	18th	X	Circular (no. 4102), earlier kiln in room 10(no. 4122) and in private house complex (no. 3896)	No. 4102: Separate hearth and firing chamber, associated with a private house, room 10, near south-eastern corner. 1.2m N-S x 1m E-W. Depth of fire pit floor 1m. Lowermost course vertical bricks. No. 4122 only 14 vertical bricks survive. No. 3896 circular, within private house, 24 bricks in barrel form, no support for kiln floor, so kiln must have been floored at higher level and stoke hole at ground.	(Borchardt, 1932) (Nicholson, 1989; 1995b)
El-Amarna	18th	X	4 pot kilns	Row against wall, joined by mud brick with vent hole leading up to each hearth. Associated with a private estate (O.49.9).	(Borchardt, 1932, pp. 73-79)
El Amarna, square Q48.4	18th	Upper stone of basalt wheel	Large Updraught kiln, smaller unfinished kiln	Rectangular enclosure with various industrial buildings	(Nicholson P. T., 1992, pp. 61-70; Rose, 1989, pp. 85-7, fig 4.2-4.4)
El-Amarna	18th	X	Pot Kiln	Associated with kiln of private house	(Peet & Woolley, 1923, p. 49)
El-Amarna, square G4, no. 2984	18th	X	Oval Kiln	2.3m N-S x 1.5m E-W x 1m deep, of which 0.75m below ground. Part of the stoke hole preserved to the south. Kiln floor half the height of the stoke hole. Area around kiln is a workshop	(Nichsolson, 1995)
El Maqata	18th	X	Small circular pot kiln	Associated with kitchen of private house south of enclosure. Lack of scale	(Davies, 1918, p. 10)
Medinet Habu	(pre-Ramesses III layer) NK	X	6 Pot kilns	To north of western fortified gate. 3 free standing, 3 joined together. Prob baking and cooking	(Hölscher, 1939, p. 73)
Mirgissa	18th dynasty	X	Area MI 6 definite, but potentially 11 others	Sizes of kilns vary; some have lowermost course vertical bricks, and engaged columns running up the walls beneath the floor, to support it.	(Vercoutter, 1970)
El Sebua (Nubia)	NK	X	Oven or Granaries	Sphinx alley of temple, to s. Square opening at ground and tapered upper	(Gauthier, 1912, p. 34)
Gurob, El Faiyoum	18th-19th	X	2 pottery kilns	Located in IA1 c40m north-east of palace, pottery kilns uncovered, together with potential pottery workshop area. Kiln 1 measure 2.8m in diameter, kiln 2, 2.4m. Workshop area possibly contains a paddling pit 1.5 x 0.95 m.	(Boatright & Hodgkinson, 2010; Hodgkinson, 2012)

Site	Period	Potter's Wheel	Kiln (type)	Other Details	Reference
Huruba, near el-Arish, Sinai	18th	X	2 Kilns	Associated with potter's workshop at location A-345. One kiln thought to be complete, (Petrie, 1894)measured 1.5m high, circumference 1.8m. Fire chamber 1m high and dug into the ground, 0.7m stoke hole faced south to avoid prevailing winds. Perforated floor preserved, 0.2-0.25m thick, holes of 0.1m diameter and spanned the kiln as a vault. Outside steps leading up to kiln. Second kiln preserved to height of 1m, fuel chamber had tiled floor, perforated floor was supported on projecting bricks.	(Oren, 1987, p. 100)
Mit Rahineh	19th Dynasty onwards, Particularly during Ramesses II	X	At least 6 kilns	Kilns cover western section of Area D3 and D4 west and extend to colonnade of the sanctuary. Possible bead factory with by-products of a glazer's workshop. Near kilns number of buildings of unbaked brick- possible storage rooms. Kiln at point 36/235 floor found to contain unbaked pottery, also shelf found pierced with round holes, resting on two vaults above the hearth upon which more pottery was placed. Kilns believed to be in use for a short time.	(Jacquet, 1965, pp. 46-59)
Near Deir el-Medina	NK or later	X	Pot Kiln	Associated with settlement 19th-22nd dyn to the left of the road from Deir el-Medina to Medinet Habu. Contained ashes, dung cakes and fragmentary bricks	Berlin Mus Inv. Nr. 23.679. Borchardt (1932, p78)
Memphis Area D (1956 excavation)	22nd (TIP)	X	6 Circular Kilns	Associated with possible potters' workshop. Kilns seem to have short use life, overlap stratigraphically, re-using bricks from previous construction.	(Anthes, 1965, pp. 22-29)
East Karnak Phase D	Late Period	X	Circular Kilns	Kiln S.P.1 faced east, preserved height of 1.38m, lined on exterior with skin of bricks, and was later filled with debris and kiln furniture and fragments of chequer (grid separating fire from vessels). Kiln S.P. 2 smaller, may in fact be an ash pit, or clay preparation area.	(Redford, 1981, pp. 14, 35)
Kom Dahab, Naukratis	Ptolemaic	X	Updraught, circular kiln	Diameter of c4.5m, furnace chamber may have stood c2.1m high. 16 wedge shaped openings survive from perforated grid.	Coulson & Leonard (1983, p66)
Tell el-Haraby	Ptolemaic	X	2 large circular updraught kilns (amphora production)	Diameter c5m. Walls of large dried bricks set into clay mortar. At top of preserved height (3m) curve inwards, suggesting dome-like roof. Bottom of firing chamber of one of the kilns found to be pierced with holes in ray-like pattern to transmit air from furnace below. Flues were reinforced by amphora stands placed inside	(Majcherek & El-Shennawi, 1992, pp. 131-2)
Athribis (Kôm Sidi Youssef)	Ptolemaic, Roman & Byzantine	X	Numerous small circular kilns	Kilns used for firing pottery, terracotta figurines and oil lamps, Layer of ash dating to reign of Ptolemy V	(Myśliwiec & Południkiewicz, 2003)
Kom Firin	2nd-3rd C CE	X	2 lime slaking kilns	Powdered lime found in both kiln and tunnel.	(Coulson & Leonard, 1983, p. 64)
Mareotis, Borg el-Arab (Along the Alexandria-Cairo Road)	Roman	X	Large kiln	Diameter c 9.6m, surrounded by vast quantities of amphora sherds. The remains of the firing chamber can be seen in ray-like pattern similar to the Tell el-Haraby example.	(Emperour & Picon, 1992)
Memphis	Roman	X	Faience kilns		(Petrie, 1911, pp. 34-37) 7)

APPENDIX I: SELECTED LIST OF KILNS, POTTER'S WHEELS AND WORKSHOPS

Site	Period	Potter's Wheel	Kiln (type)	Other Details	Reference
Dakhla Oasis	Roman (1st -3rd C CE)		Circular updraught kilns	Much larger than OK examples, diameter 1.4m-2.27m. 4 distinct types of kiln, though all are circular, with thick walls. The firing chamber is located in a pit cut into the ground, several contained unfired pottery fragments. Site 33/390-19-3 and 33/390-K9-I.	Hope (1979, pp. 123-127)
Saqqara (workshops of monastery of St Jeremia)	Older than 6th C CE	X	6 kilns (Nr. 114, 116, 117, 118, 121, 126)	Set in a complex of earlier abandoned buildings. Varied as to function and size. Kiln 114- floor was dug and walls constructed with half bricks, E part was left open for fuel. Kiln 116- constructed of 3 round walls (c.1.20m) inner wall of fired bricks, outer mudbrick wall separated by gap of 10cm and space filled with sand. Used to consolidate kiln. 2 openings, one to W, one to E. Kiln 117- built of mudbrick (c1m) red fired on the inner face, contains mostly burnt lime, shape differs from classical type of pottery kiln, hole in the ground consolidated with mudbrick. Prob used for burning limestone. Kiln 118- mudbrick (c1.2m) large quantity of painted sherds came from debris, poss used as storage area. Kiln 121- set in small room in earlier complex. Doorway was used to insert arched stokehole. Kiln fill consisted of dark sandy material including bricks (destruction layer) then layer of greenish-yellow sand with rubble of flooring of the kiln, including amphora used to fill in holes of perforated floor. Brick ledge where floor rested on still preserved in places. 10 sand filled trapezoidal holes set into the ledge, prob ventilation holes 0.12 x 0.27m on kiln face, upper opening c 0.15-0.25m, 10 of which survive at 23cm intervals. Allowed hot air through fuel room to vessel stack in same way as perforated floor. Height of kiln about 4m, internal diameter 2.6m. Large variety of sherds found surrounding this.	(Ghaly, 1992, pp. 161-163)
Tod in the Valley of the Queens	10th-11th C CE	X	2 Circular Kilns	2 kilns of fired mudbrick, very levelled out so difficult to determine what kiln was used for as there is an absence of associated materials.	(Lecuyot & Perrat, 1992, pp. 176-177)

	Nile A	Nile B1	Nile B2	Nile C	Nile D	Nile E
Ground-mass	Fine, silty clay	Silty	Silty	Fairly silty	Silt with lots limestone	Silt with lots inclusions
Inclusions	Fine sand, mica	Abundant fine sand, mica	Larger & more particles sand, mica, limestone	Variable sand, med limestone, mica, grog, rock particles	Lots limestone	Lots fine/med/ coarse sand, mica
Organics	Fine straw	Straw <2mm	Lots fine-med straw, coarse straw	Predom straw particles	Lots fine straw	-
Fracture Colour	Brown/greyish brown	Reddish Brown	Brown, narrow red or black cores	Grey-brown	Outer zones of red and violet	Brown/ black
Firing Temp °C	700-750	750-850	500-800	500-800	750-850	500-800
Porosity	Moderate	Moderate	Loose/open- moderate	Open	Loose-mod	Open
Hardness	medium	Soft-med	Med-hard	Soft-med	Hard	Soft
Transverse Strength	Medium	Low-medium	Med-great	Very low-med	Great	Low-med
Examples	Black topped red ware UC5688	Carinated cups UC 20500, UC17855	Bowls, Flask, spouted jars eg UC30223, UC18002	Jars and pot stands Fitz Mus E250.1899, UC17988	Amphora SIP used for handmade storage jars	Cooking vessels and bread moulds
Use	Badarian funerary equipment	Common from OK, drinking cups, fine wares	Common in all periods and regions, usually for handmade vessels	Containers with thick walls. All periods and regions	Large vessels with thick walls	Restricted geo limits, E Delta, Memphite region to Fayoum. Bread moulds and cooking pots MK-NK

Appendix II: The Vienna System of Clay Fabrics: The Nile Silts (After Arnold & Bourriau, 1993, pp. 170-175)

Appendix II: The Vienna System of Clay Fabrics: The Nile Silts

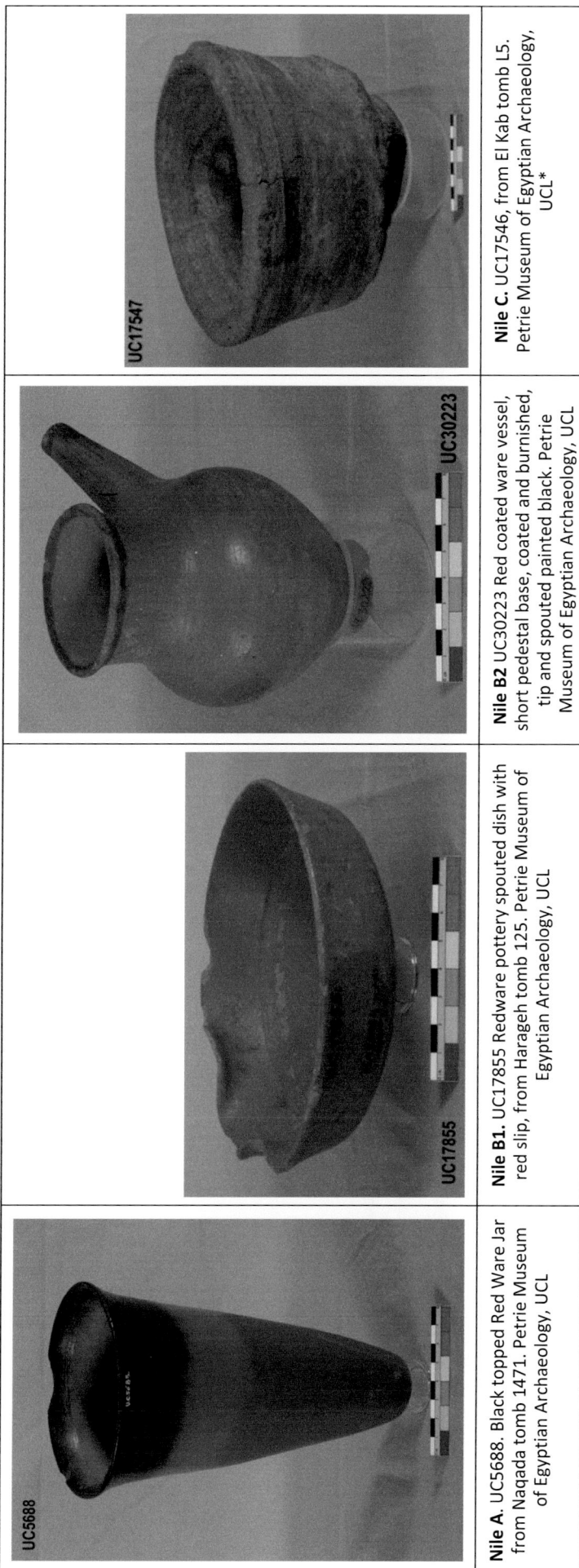

Nile A. UC5688. Black topped Red Ware Jar from Naqada tomb 1471. Petrie Museum of Egyptian Archaeology, UCL

Nile B1. UC17855 Redware pottery spouted dish with red slip, from Harageh tomb 125. Petrie Museum of Egyptian Archaeology, UCL

Nile B2 UC30223 Red coated ware vessel, short pedestal base, coated and burnished, tip and spouted painted black. Petrie Museum of Egyptian Archaeology, UCL

Nile C. UC17546, from El Kab tomb L5. Petrie Museum of Egyptian Archaeology, UCL*

* Nile C inclusions- straw survives as carbonised particles, as white or grey silica skeletons of its cellular structure and as impressions in the paste. Rod shaped particles usually becomes orientated parallel to the vessel wall. In the case of the vessels thrown on the wheel or shaped on a turning advice. Necessary to examine a sherd fractured parallel to the throwing lines, in order to identify the shape and frequency of the straw particles (Arnold & Bourriau, 1993, p. 173).

	Marl A1	Marl A2	Marl A3	Marl A4
Groundmass	Fine and homogenous calcareous clay	Fine and temper evenly scattered throughout paste	Closest to modern Qena clay, very fine, perhaps naturally levigated	Coarsest texture of the marl As
Inclusions	Fine-med crushed limestone. Clearly visible. Fine sand, mica	Fine sand and limestone particles and unmixed marl, mica	Few mineral inclusions, prob not added	Greatest quantity of fine to coarse sand inclusions, scattered mica
Organics	Few pieces of straw	-	Occasional straw particles	Some conspicuous straw
Fracture Colour	Pale-light red	Pale red, yellow to grey-white surface and fracture is pale red	Pale greenish grey, sometimes pink spots or zones. Surface pale yellow to reddish yellow	Considerable range of colour Light red and greenish grey
Firing Temp °C	800	1000	1000	800-1000 If fired between 800-850, strong reaction to hydrochloric acid, at higher temps no reaction
Porosity	Dense, elongated voids	Dense without conspicuous pores	Elongated, roughly rectangular pores, but otherwise dense	Open texture due to burnt out limestone
Hardness	Hard and firm	Extremely hard	Medium-firm	Varies crumbly & soft to hard/firm
Transverse Strength	Great, breaks sharply when struck	Great, surface feels distinctly grainy	Great	Medium-great
Examples	Meidum ware bowls and Petrie's D-ware.	Squat jar from Hiw, Fitz E.63.1899	Squat Jar UC18385 (12th dynasty)	Storage Jar, Hu (Diaspolis Parva) Fitz E.202.1899
Use	Common from Naq II to Ok	Occurs from MK, most common in SIP, more plentiful in Upper than Lower Egypt	Early MK-NK, seems to originate in Upper Egypt where it is common	Occurs in MK, most common NK

Appendix III: The Vienna System of Clay Fabrics: The Marls (After: (Arnold & Bourriau, 1993, pp. 176-178))

APPENDIX III: THE VIENNA SYSTEM OF CLAY FABRICS: THE MARLS

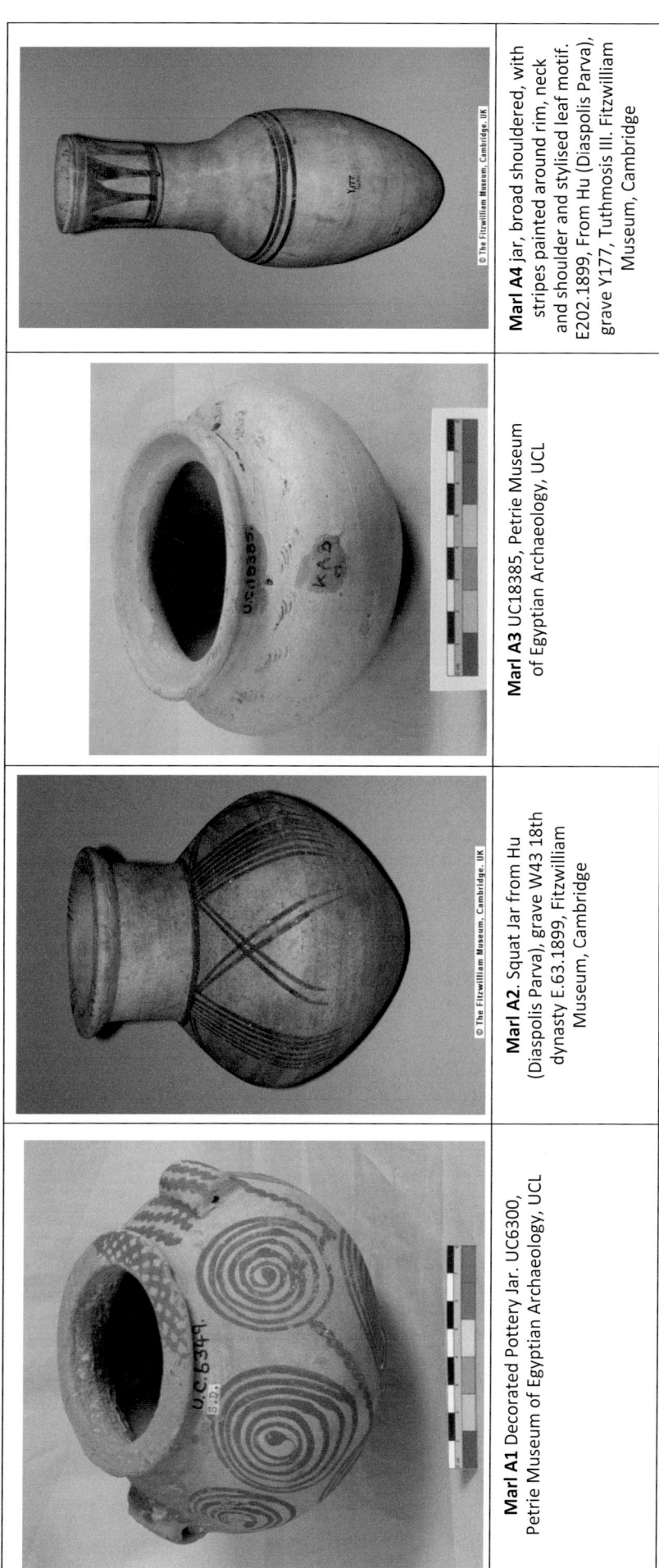

Marl A1 Decorated Pottery Jar. UC6300, Petrie Museum of Egyptian Archaeology, UCL

Marl A2. Squat Jar from Hu (Diaspolis Parva), grave W43 18th dynasty E.63.1899, Fitzwilliam Museum, Cambridge

Marl A3 UC18385, Petrie Museum of Egyptian Archaeology, UCL

Marl A4 jar, broad shouldered, with stripes painted around rim, neck and shoulder and stylised leaf motif. E202.1899, From Hu (Diaspolis Parva), grave Y177, Tuthmosis III. Fitzwilliam Museum, Cambridge

	Marl B	Marl C1	Marl C2	Marl D	Marl E
Groundmass	Homogenous and dense	Homogenous and dense	Homogenous and dense	Fine and homogenous	Dense
Inclusions	Abundant sand, c40% of the paste, added as temper, coarse and sub angular, some mica	Mass of fine and medium decomposed particles of limestone. Large quantities of fine and medium sand	Limestone particles remain intact after firing. Sand is present in larger quantities than limestone.	Limestone particles added as temper. Smaller in size than Marl C, varying from fine to coarse, up to 25% of the fabric. Fine to coarse sand, mica and dark rock material. Surface feels gritty.	Similar to Marl B, abundant medium to coarse sand, mica and unmixed groundmass material
Organics	Occasional straw	-	-	Rare	Abundant particles of medium to coarse straw added to paste
Fracture Colour	Core is usually pink, outer zones of grey-white to green	Fracture almost always zoned, red with a grey or black core, sometimes vitrification	Uniform colour ranging from red to brown	Surface light green to grey, fracture pale greyish brown, higher fired e.g.s show red/brown outer zones or unified red-brown core.	Surface yellowish white to greenish grey and in fracture from pink to greenish grey. Green highly fired examples do not react to HCl
Firing Temp °C	Over 800	850-1000	750-850	850-1000	
Porosity	Dense, sometimes with vitrification	Dense	Dense	Dense	Open, porous texture dominated by voids
Hardness	Hard and firm, but sand can make it crumbly if overfired	Hard and firm	Hard and Firm	Hard and Firm	Hard
Transverse Strength	Low-medium	Low-medium	Low- Medium	Great	Great
Examples	Storage pots E.161.1902	Cooking pot UC18636		19th dynasty Amphorae such as handles in Fitz museum EGA.4157-8.1943. Stamps royal names on handles.	
Use	Large-med storage vessels. Occurs in SIP to 18th dynasty, most common in Upper Egypt. May be imported from north into south. Plentiful in Deir el-Ballas, rare in Memphis	Common in the Memphite – Faiyum region	Variant of Marl C1	Common in 18th -19th dynasties in Delta and Memphis-Faiyum region, seems to occur in south only as northern imports. E.g. wine amphora	Commonly used for thick-walled vessels such as bread trays. Deliberate addition of straw related to function. Seems to occur short time SIP-18th and origins in Ballas

APPENDIX III: THE VIENNA SYSTEM OF CLAY FABRICS: THE MARLS

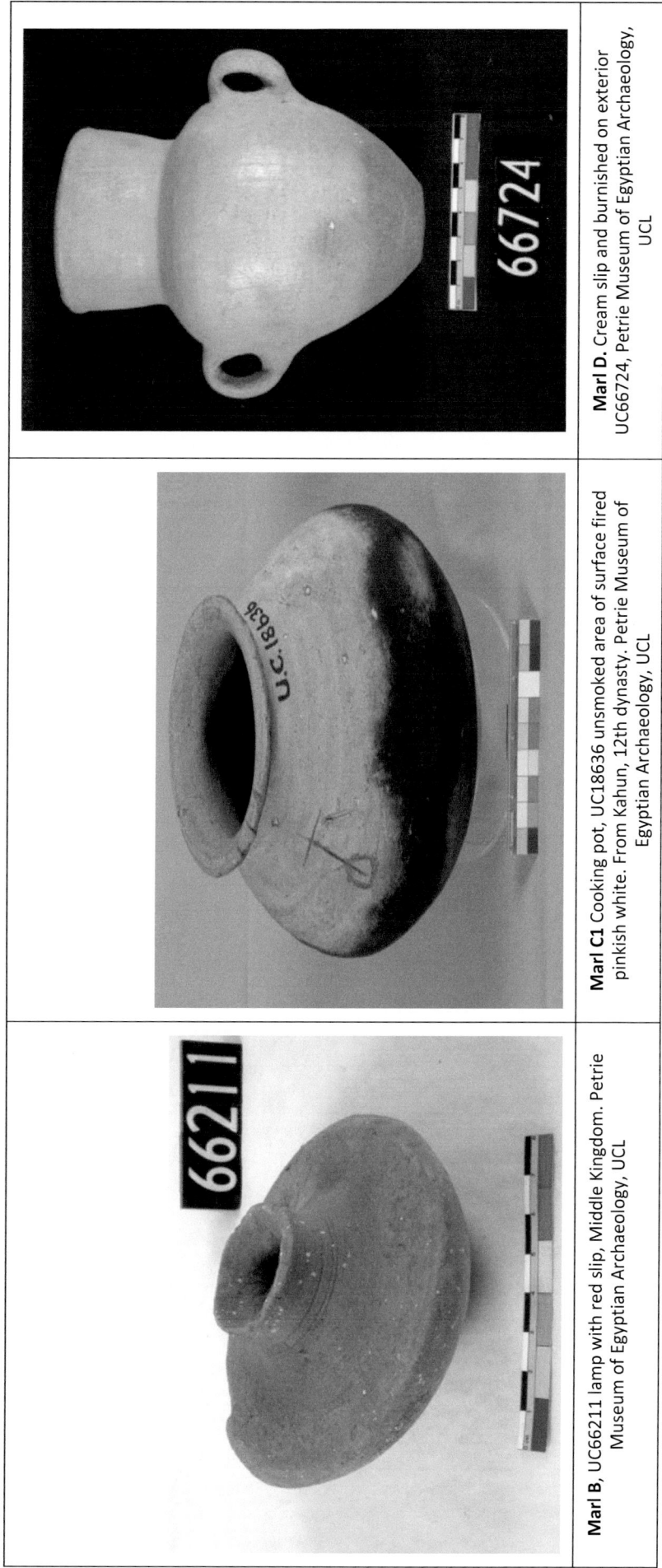

Marl D. Cream slip and burnished on exterior UC66724, Petrie Museum of Egyptian Archaeology, UCL

Marl C1 Cooking pot, UC18636 unsmoked area of surface fired pinkish white. From Kahun, 12th dynasty. Petrie Museum of Egyptian Archaeology, UCL

Marl B, UC66211 lamp with red slip, Middle Kingdom. Petrie Museum of Egyptian Archaeology, UCL

Appendix IV: The Concrete (Drawing: A. Davies) and Granite Wheel Bearings (Drawing S. M. McConnell & Sons) Plans based on BM32622

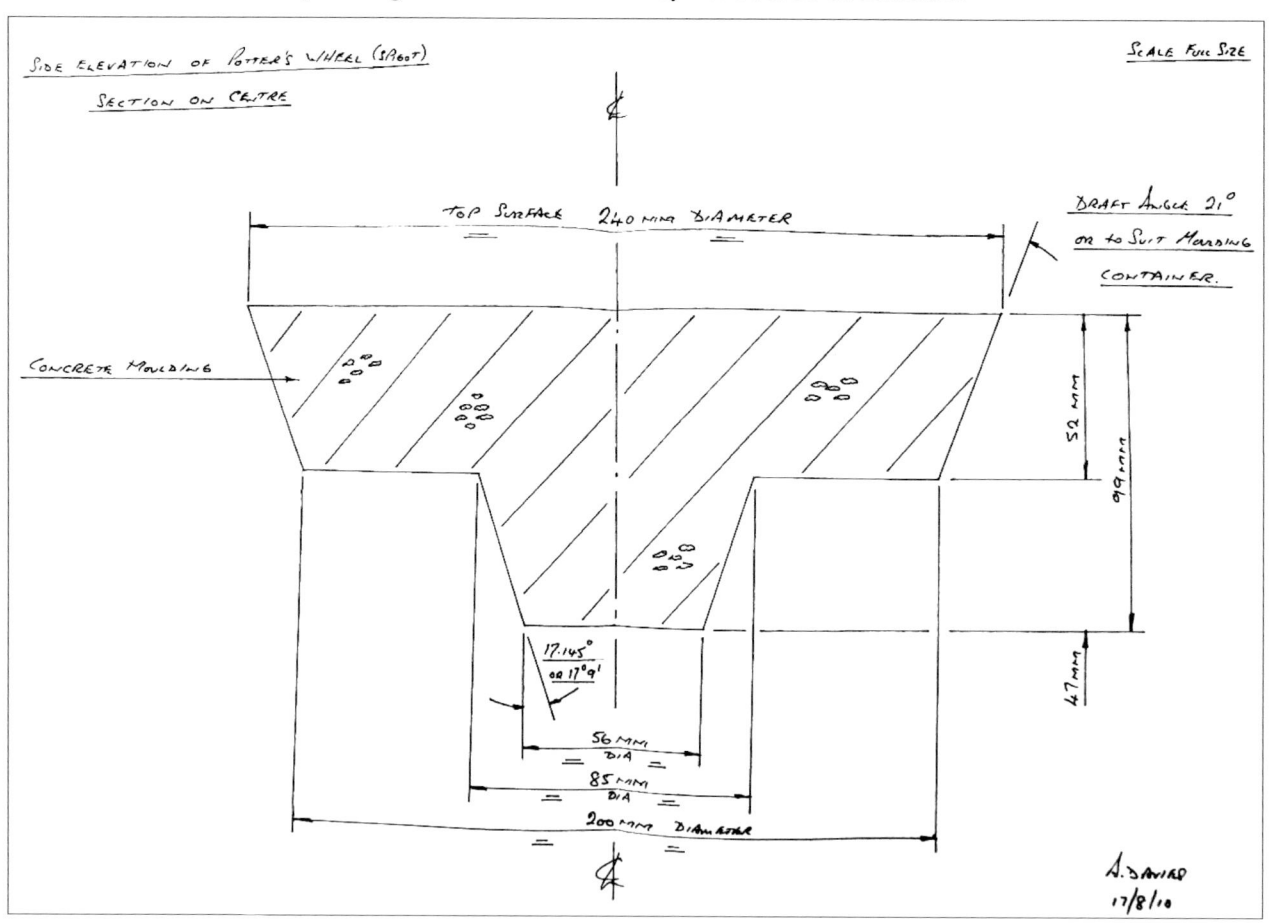

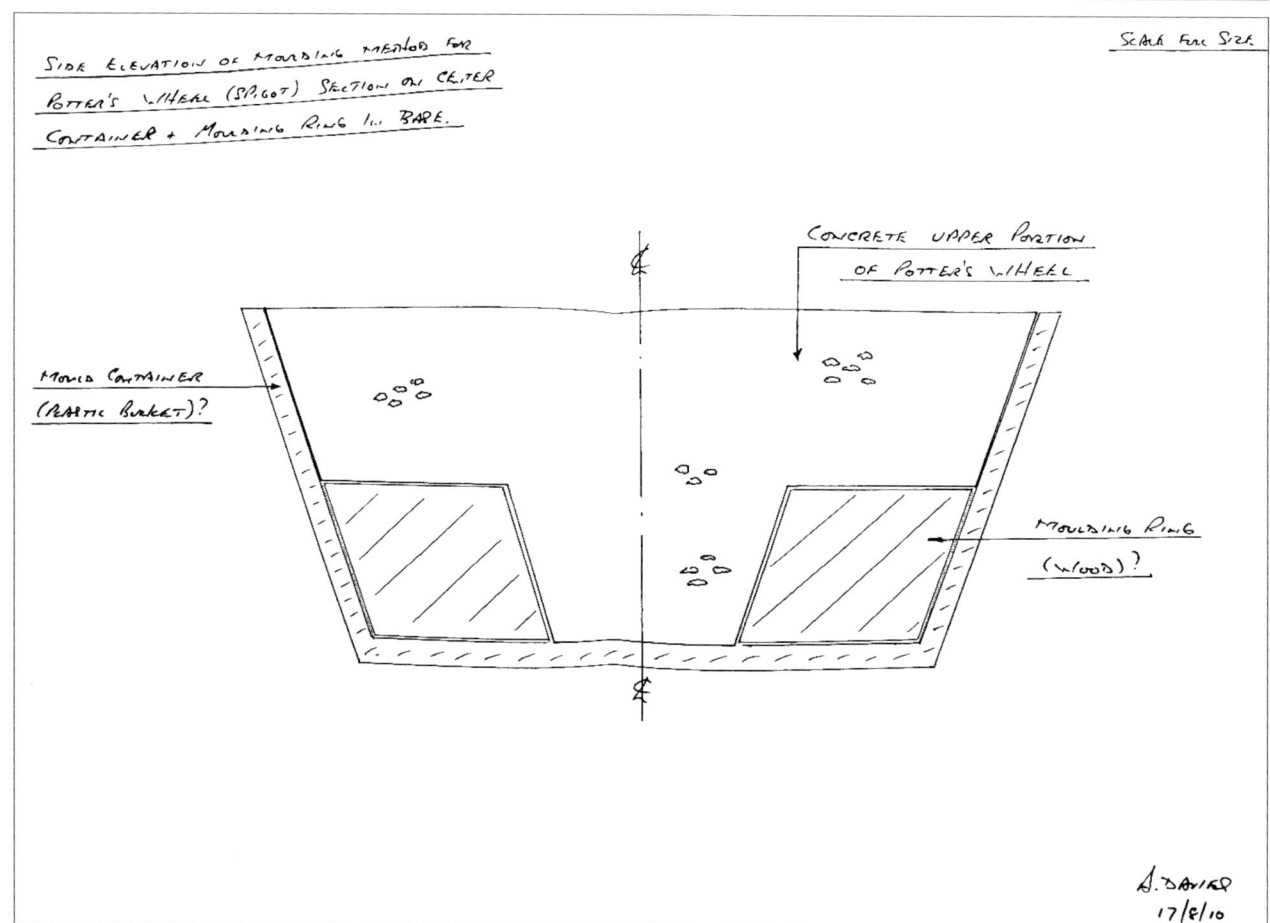

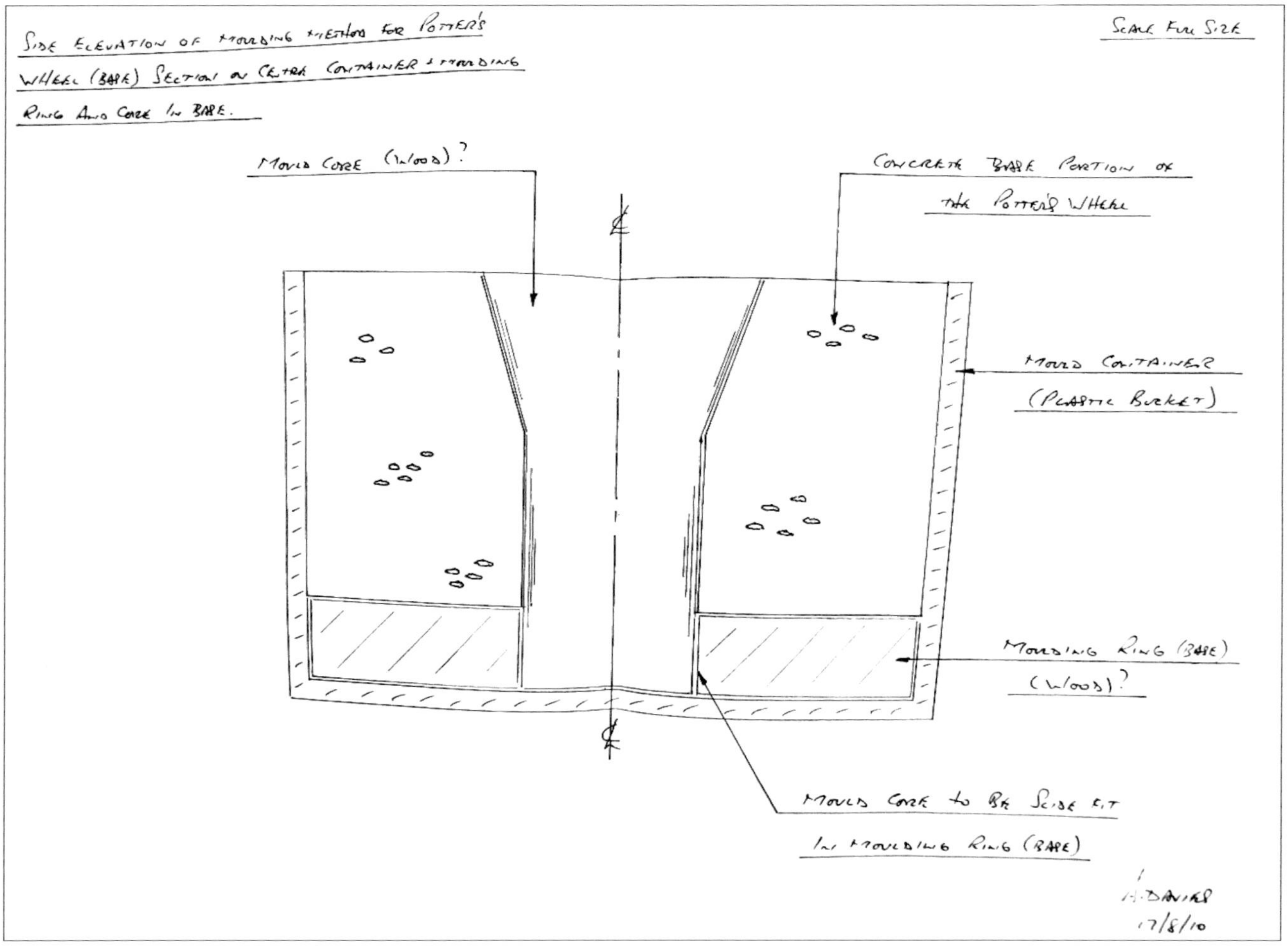

APPENDIX IV: THE CONCRETE AND GRANITE WHEEL BEARINGS PLANS BASED ON BM32622

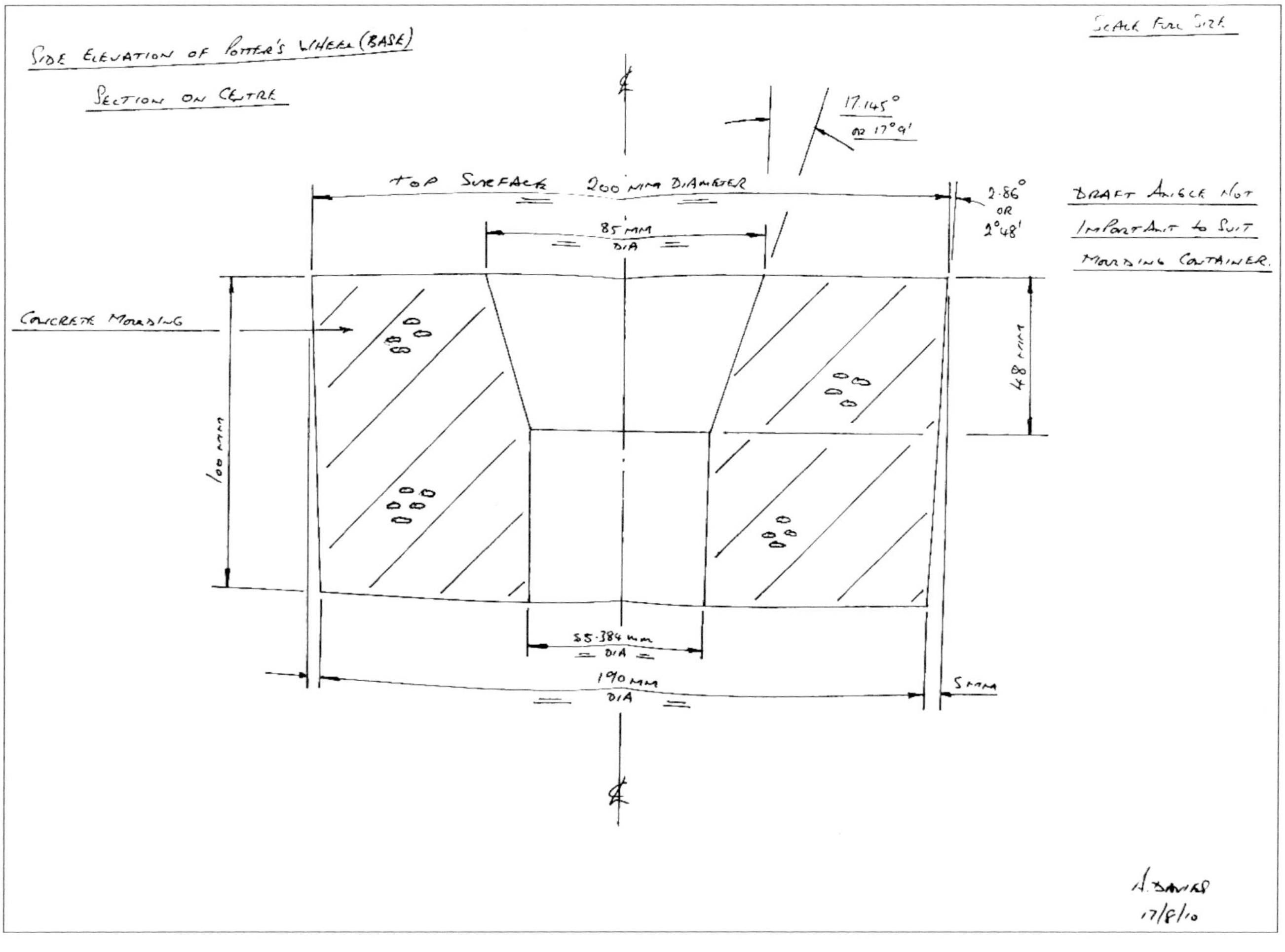

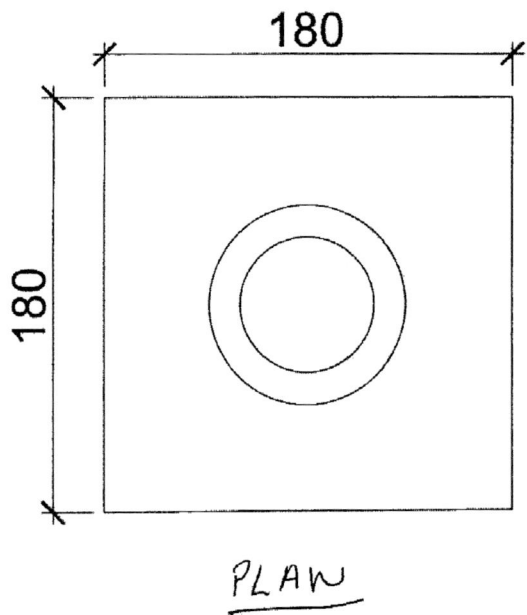

PLAN

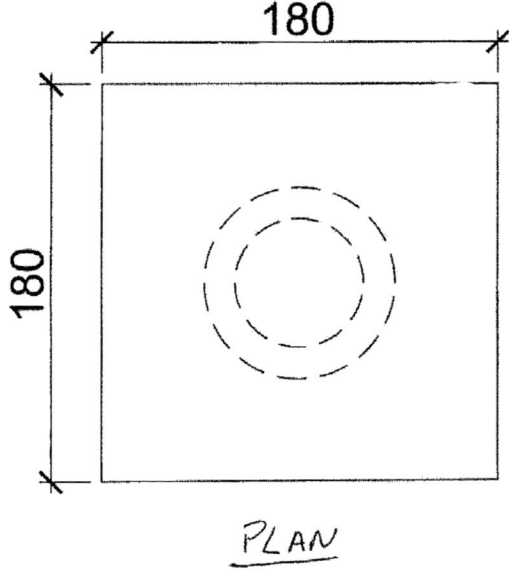

PLAN

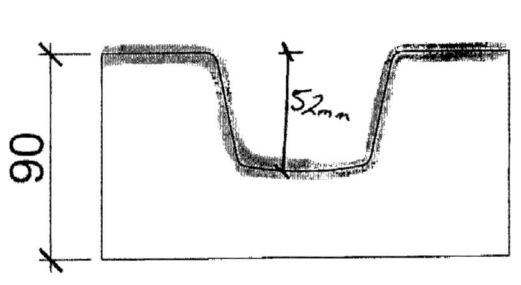

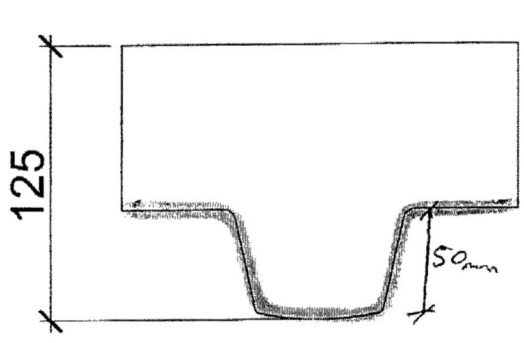

Grinding Wheel Bearing Stones

Honed Faces to allow Stones to Spin

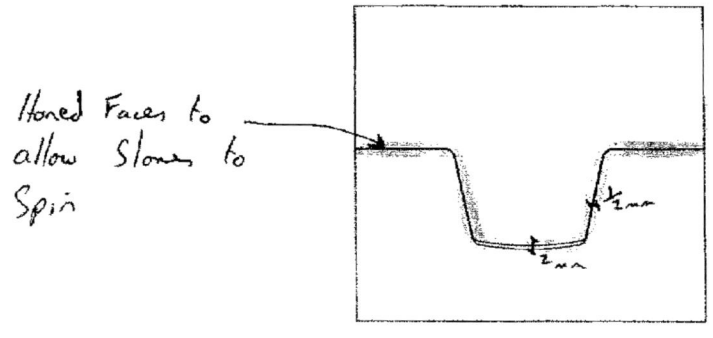

Pivot Stone Needs to Fit in Socket Stone with 2mm Gap at Bottom & ½mm Each Side